THE GOLDEN SECTION

THE GOLDEN SECTION

An Ancient Egyptian and Grecian Proportion

Steven L. Griffing

Library of Congress Control Number:		2006909292
ISBN:	Hardcover	978-1-4257-2951-6
	Softcover	978-1-4257-2950-9

To order additional copies of this book, contact:
Xlibris Corporation
1-888-795-4274
www.Xlibris.com
Orders@Xlibris.com
31020

ACKNOWLEDGEMENTS

In Chapter I., Section F., Figure 9 is reproduced with permission from *Webster's Third New International® Dictionary, Unabridged* ©1993 by Merriam-Webster, Incorporated (www.Merriam-Webster.com). Figures 10, 11, 12, 14, and 15 are reproduced from *Men of Modern Mathematics* (A history chart of mathematicians from 1000 to 1900), ©1966, with permission from the International Business Machines Corporation, Somers, New York, and are held by IBM Corporate Archives. Figure 16 is reproduced from *The Diagonal* by Jay Hambidge, ©1920, Vol. I, No. 3 (January, 1920), p. 55, with permission from the Hambidge Foundation in Rabun Gap, Georgia. Figure 17 is reproduced from *Dreams* by C. G. Jung, ©1974, p. 199, with permission from the Princeton University Press, Princeton, New Jersey.

I am indebted to Dr. Thomas Schwartzbauer, a former professor in the mathematics department of the Ohio State University, for the information and help with questions, he gave me, regarding the Pythagorean musical scale (Chapter I., Section D., 2.). I had two meetings with him at O. S. U. in the first half of 1986. Credit, also, goes to Tom Boginksy, a German language student at the Ohio State University, for helping me translate from German, parts from Ernst Moessel's book, *Die Proportion In Antike und Mittelalter*, ©1926, dealing with the rock tomb at Mira, the Cathedral of Notre Dame, and the Palace of the Alhambra (Chapter VI., Sections C., E., and F.). I had two meetings with him at O. S. U. on 05/29/86 and 06/05/86.

In Chapter VI., Section B., Figures 106 and 107 are reproduced from *The Diagonal* by Jay Hambidge, ©1920, Vol. I, No. 3 (January, 1920), pp. 50 and 51, with permission from the Hambidge Foundation in Rabun Gap, Georgia. In Section D., Figures 108 and 109 are reproduced from *The Parthenon and Other Greek Temples* by Jay Hambidge, ©1924, pp. XVII. (Preface) and 10, with permission from the Hambidge Foundation in Rabun Gap, Georgia.

CONTENTS

INTRODUCTION

We all know that the aspiration to achieve beauty is a very important aspiration of man, just as the aspiration to achieve human beauty has been very important to the human species in the one and one half to four million years of its evolution. It is because of the natural rhythm or dynamic symmetry of the human body that the human body is so beautiful. The dynamic symmetry of the human body and of many other animals is based on the golden section. The human body and the bodies of many other animals are living proof to the fact that the golden section is, naturally, beautiful, or pleasing to the human eye in the human body, and to the eyes of many other animals in the bodies of these animals. So, also, is the golden section, naturally, beautiful when reproduced in the art (including drawing, painting, pottery, and sculpture), architecture, and materials' design of man.

If one examines the cultural heritages of all races of people, and nations, or tribes of races of people on planet earth, it can be, easily, deduced that the fundamental principal running throughout is that of the achievement, and/or magnification of natural beauty. It is in the golden civilizations of Egypt and Greece, that this ideal of the achievement of natural beauty in the reproductive arts reached a peak. Egypt and Greece were the only two nations to ever use the golden section to any extent. During the great, classical ages of these two nations, the golden section was used extensively.

CHAPTER I.

The Historical Development of the Golden Section from Ancient Egypt

I

The Historical Development of the Golden Section from Ancient Egypt

A. Introduction

The golden section is an ancient Egyptian and Grecian proportion. It was used as early as in the building of the Great Pyramid of Cheops by the Egyptians around 2500 B.C. The Greeks adopted the ideas of the golden section from the Egyptians around the sixth century B.C.

I am indebted to Jay Hambidge for the following historical account, as to the method of development, and origin of the golden section in ancient Egypt, India, and Greece.

It is impossible to use dynamic symmetry unconsciously. Curiously, there were but two peoples who did use dynamic symmetry, the Egyptians and the Greeks. It was developed by the former very early as an empiric or rule-of-thumb method of surveying. Possibly the date is as early as the first or second dynasty. Later it was taken over as a means of plan making in architecture and design in general. The Egyptians seemed to attach some sort of ritualistic significance to the idea as it is found used in this sense in temple and tomb, particularly in the bas-reliefs which were used so plentifully to adorn these. It is also curious that the Hindus, about the fifth or eighth century B.C., possessed a slight knowledge of dynamic symmetry. A few of the dynamic shapes were actually worked out and appear in the Sulvasutra, literally "the rules of the cord," and were part of a sacrificial altar ritual. But to what extent it may have been used in Hindu art is not known, because examples containing its presence have disappeared. The Greeks obtained knowledge of dynamic symmetry from the Egyptians some time during the sixth century B.C. It supplanted, probably rapidly, a sophisticated type of static symmetry then in general use. In Greece, as in India and in Egypt, the scheme was connected with altar ritual. Witness the Delian or the Duplication of the Cube problem. The Greeks, however, soon far outstripped their Egyptian masters and, within a few years after acquiring the knowledge, apparently made the astounding discovery that this symmetry was the symmetry of growth in man. (1919, p. 2)

The golden section was discovered by the Egyptians, and has been used in art and architecture, most commonly, during the classical ages of Egypt and Greece. Its basic importance is derived from the natural proportions of the bones of the human body. It is a proportion, which, besides being used for natural beauty in art, has fundamental geometric properties, which will, also, be investigated in this book.

The formula for the golden section is derived from a line segment, but its applications in geometry and art are almost limitless. It is found in such fundamental geometric structures as the golden triangle, the golden rectangle, and the pentagram. If a line segment is divided into two unequal segments, such that the ratio of the lengths of the whole to the longer segment is equal to the ratio of the lengths of the longer segment to the shorter segment, then this line segment is said to be divided into golden section. This ratio is called the golden ratio, and will be denoted by τ in this book. This ratio can be found as follows. Consider the diagram in Figure 1.

Figure 1

$$(X + Y)/Y = Y/X$$
$$X(X + Y) = Y^2$$
$$X^2 + XY = Y^2$$
$$1 + Y/X = (Y/X)^2$$
$$(Y/X)^2 - Y/X - 1 = 0$$
$$\therefore \quad (Y/X) = (1 \pm \sqrt{(1+4)})/2$$
$$= (1 \pm \sqrt{5})/2$$

Because the value of the ratio of two positive lengths can only be positive, we have to take the positive root. Therefore,

$$Y/X = (1+\sqrt{5})/2$$
$$= 1.618033989\ldots$$
$$= \tau$$

The golden ratio is an irrational number, and written in decimal form is a non terminating decimal.

The continued fraction for the golden ratio is

$$\tau = 1 + \cfrac{1}{1 + \cfrac{1}{1 + \cfrac{1}{1 + \cfrac{1}{1 + \cfrac{}{\ddots}}}}}$$

Because this continued fraction consists of all ones, it is "The simplest of all infinite simple continued fractions" (Olds, 1963, p. 81). "The convergents to τ are 1/1, 2/1, 3/2, 5/3, 8/5, 13/8, ..., both numerators and denominators being formed from the sequence of integers 1, 1, 2, 3, 5, 8, 13, 21, 34, ..." (Ibid.) This is the Fibonacci number sequence.

The Fibonacci numbers are an additive series of numbers, which are found commonly in nature, especially, in plant morphology in a property called phyllotaxis, or the method of the geometric arrangement of leaves on a stem, or branch. The Egyptians had discovered this property, long ago. The ratios of the larger to the smaller of two consecutive Fibonacci numbers forms a very close approximation to the golden ratio, and, gradually, approaches it, as the Fibonacci numbers grow infinite. This is how the golden section is found applied to the architecture of plants.

The Egyptians applied the ratio of the larger to the smaller of two of the Fibonacci numbers, lower in the series, to their art and architecture. This includes in drawing, design, bas-relief wall carving, painting, pottery, sculpture, and even in tomb furniture. The properties of the Fibonacci numbers and their use by the Egyptians will be explored later in this chapter.

The first terminology used for the golden section, known to modern scholars, was that of Euclid. In his work, *The Elements*, he used the terms "To cut a given finite straight line in extreme and mean ratio" (Sir Thomas Heath, 1956, Book VI., Proposition 30). This can be interpreted to mean the "division of a line segment into two unequal parts such that the ratio of the whole to the larger part is equal to the ratio of the larger to the smaller" (H. S. M. Coxeter, 1953, p. 135).

In modern terminology with any pair of equivalent fractions, the two outside terms are called the extremes, while the the two inside terms are called the means (see Figure 2). Therefore, it seems appropriate that Euclid should use the terms extreme and mean ratio for the division of a line segment by one point. If this point were placed in such a way, that the ratio of the lengths of the whole to the longer segment would be equal to the to the ratio of the lengths of the longer segment to the shorter segment, then a pair of equivalent fractions would be formed. This line segment could be said to be divided into extreme and mean ratio. Then, this line segment could be said to be divided into golden section.

$$W/X = Y/Z$$
W and Z are called the extremes.
X and Y are called the means.

Figure 2

The next terminology used for the golden section, known by modern scholars, was that used by Proclus. 'The Greek geometers of the Platonic school called it ἡ τομή, "the section" *par excellence*, as reported by Proclus (*On Euclid*); Luca Pacioli, Leonardo's friend, called it "The Divine Proportion."' (Ghyka, 1977, p. 4)

> Proclus (410-483 A.D.) writes: "Eudoxus of Cnidos . . . an associate of Plato's school, . . . to the 3 proportions added another three . . . and increased the number of theorems about *the section* (περὶ τὴν τομὴν) which had their origin with Plato" (*On Euclid*).
> Bretschneider identifies ἡ τομή with the Golden Section (which in German is also called *the* continuous proportion, *die stetige Proportion*). (Ibid., p. 125)

The golden section was discovered by the Greeks about the middle of the 5th century B.C., probably from the regular pentagon: its diagonals form the pentagram, which contains 200 golden ratios. The pentagram goes back to Babylonian culture; the Pythagoreans used it as a sign of salvation and secrecy. The Greek term for the golden section is "the division of a line in extreme and mean proportion." The Italian mathematician Luca Pacioli called it *divina proportione* (divine proportion); the German Johannes Kepler (d. 1630) *sectiò divina* (divine section), for to him it symbolized the Creator's intention "to create like from like." The term *proportio continua* appears in the 16th century; probably by translation this became "continuous (*i.e.,* uninterrupted) division," in the 18th century. The term "golden section" first appears in 1830. (Menninger, 1972, vol. 10, p. 542)

The golden section has still been used in modern day, and has, since the classical age of Greece, by such great artists and architects as Seurat, Mondrian, and Le Corbusier. However, it has been used much less, commonly, by the European peoples than it was used, during the classical ages of Egypt and Greece.

I

The Historical Development of the Golden Section from Ancient Egypt

B. The Fibonacci Numbers

1. Introduction

The Fibonacci numbers are an additive series. They are formed by the equation

$$u_1 + u_2 = u_3,$$

where the first two numbers of the series are one and one. There are many additive series of this nature. But, the Fibonacci numbers are considered important, because they are found in nature, especially in the property called phyllotaxis, and are thought to have origins in the microscopic structure of genes.

The Fibonacci numbers were first discovered by Leonardo of Pisa, better known as Fibonacci, and were published in his book, *Liber Abaci*, in 1202 A.D. Fibonacci means son of a simpleton, or son of an ass. In this book, Fibonacci proposed a problem in which each year the number of rabbits produced will be equal to the sum of the number of rabbits produced in the respective two previous years. Therefore, the sum of rabbits in each successive year will be equal to another Fibonacci number.

Fibonacci was of Italian descent, and was raised in northern Africa. He was trained in mathematics by the Muslims of the Barbary coast. According to Peter Tompkins, Fibonacci had traveled with his father across northern Africa to Egypt. Fibonacci introduced the Hindu-Arabic numerals to western Europe.

As with all additive series of this nature, the ratio of the larger to the smaller of two consecutive Fibonacci numbers approximates the golden ratio. For instance, 13/8 is 1.625; 8/5 is 1.6; and 5/3 is $1.\overline{6}$. As the size of the Fibonacci numbers approaches infinite, the ratio of the larger to the smaller of two consecutive Fibonacci numbers approaches the golden ratio.

It can be, easily, shown that for the continued fraction

$$X = 1 + \cfrac{1}{1 + \cfrac{1}{1 + \cfrac{1}{1 + \cfrac{1}{1 + \cdots}}}}$$

$X = \tau$, or the golden ratio. This can be proven, as follows.

$$X = 1 + 1/X$$
$$\therefore \quad X^2 = X + 1$$
$$\therefore \quad X^2 - X - 1 = 0$$
$$\Rightarrow \quad X = (1 \pm \sqrt{(1+4)})/2$$
$$= (1 \pm \sqrt{5})/2$$
$$= \tau$$

See Section A., Chapter I.

In number theory, the Fibonacci numbers have many interesting properties. One of these is that every third number is an even number. This is so, because the sum of two odd numbers is, always, an even number. The sum of an odd and an even number is, always, another odd number. We start with one and one. One is the first odd number.

Except for one, five is the only number in the Fibonacci number series, which numerates itself in the order of the series. Also, it is the third prime number. Five and three are two consecutive Fibonacci numbers. 144 is the twelfth Fibonacci number (F_{12}), and is equal to twelve squared!

I

The Historical Development of the Golden Section from Ancient Egypt

B. The Fibonacci Numbers

2. Phyllotaxis

According to the *Webster's New World Dictionary of the American Language* (1957), phyllotaxy is "the arrangement of leaves on a stem", or "the principles of such arrangement" (s. v., "phyllotaxy"). The phyllotaxis of almost any green plant contains the Fibonacci numbers, and can be checked as follows: by starting at the base of a stem and counting the number of leaves above the first leaf, but not including the first leaf, until we arrive at a leaf directly above the first leaf, we arrive at a Fibonacci number. Also, by counting the number of times we circle the stem before arriving at a leaf directly above the first leaf, we arrive at another Fibonacci number. The ratio of the latter to the former of these two Fibonacci numbers, usually, gives us the ratio of the smaller to the larger of two alternate Fibonacci numbers. This ratio tells us the degree of turn of a full circle between consecutive leaves around the stem of a plant. A series of such fractions is listed in Table 1.

Table 1

$$\frac{F_n/F_{n+2}}{}$$

½
1/3
2/5
3/8
5/13
8/21
13/34

•

•

•

For instance, the degree of turn of a full circle between the consecutive leaves on a stem of an elm tree is ½. The degree of turn of a full circle between the consecutive leaves on a stem of a beech tree is 1/3. The degree of turn of a full circle between the consecutive leaves on a stem of an oak tree is 2/5, etc.

The ratio of the smaller to the larger of two alternate Fibonacci numbers approaches the value of τ^2, as the Fibonacci numbers approach infinite. Therefore, the values of these fractions all approximate the value of τ^2.

Patterns of three in nature include the three-leafed clover, the six-petaled daffodil, the six-petaled amaryllis flower, the flowers of other bulb plants, and the twelve-petaled flower of the holly plant. The daffodil flower has two sets of three petals each, which overlay each other. The amaryllis flower has two sets of three petals each, which overlay each other. Also, the flower head of an amaryllis plant has six stamens, bearing pollen, and one style with three stigmas, extending from the center of the flower, which receives the pollen to produce seed. And, the small, yellowish flower of the holly plant has two major layers of six petals each, each consisting of two layers of three petals, which overlay each other.

Patterns of five in nature include in leaves and in petals. The leaves of many different types of deciduous trees have five major veins. Examples of flower plants with five-petaled flowers are the California poppy, buttercup, blue flax, swamp rose mallow (Hibiscus), pansy, rhododendron, primrose, Salpiglossis (of the nightshade family), pumpkin flower, morning glory, and harebell (of the bellflower family). Examples of trees with five-petaled flowers are the apple and orange trees. Many of these flower heads have five sepals at the base of the flower head, between which the petals are produced. Many of these have a style with five stigmas.

Many flower heads are either five or six-petaled. Most of these six-petaled flower heads contain two sets of three petals each, which overlay each other. Similarly, many of the four-petaled flower heads contain two sets of two petals each, which overlay each other. So, in almost all flower heads, some form of phyllotaxis can be found. Therefore, plants with flower heads containing 1, 2, 3, 4 (or 2*2), 5, or 6 (or 3*3) petals can all be found to contain the Fibonacci numbers in the geometry of their design, or basic genetic structures. This would make sense, because the flower head is the part of the plant, which must attract insects for pollination, and, therefore, must be the most beautiful.

Also, the flower heads of many plants have florets in the center of the flower head, that are arranged in the form of logarithmic spirals. These logarithmic spirals often contain a number of florets, which is an exact Fibonacci number. These plants include the sunflower and the daisy.

In a pineapple, one can find the ripened ovaries forming three series of logarithmic spirals, running in different directions. In one direction, lie five rows of logarithmic spirals. In another direction, lie eight rows of logarithmic spirals. And, in another direction, lie thirteen rows of logarithmic spirals. Similarly, in pine cones, one can find the scales forming five rows of logarithmic spirals in one direction, eight rows of logarithmic spirals in another direction, and, sometimes, in larger pine cones, thirteen rows of logarithmic spirals in another direction.

More importantly, is the meristem tissue growth at the bud of a pine tree branch. Its helical whorls, or logarithmic spirals resemble, very much, the logarithmic spirals of a pine cone, as does its phyllotaxy. In the meristem tissue growth at the bud of a pine tree branch, eight spirals can often be counted going one way, and thirteen another way. This meristem tissue growth shows us the entire map, or network design of its whole future growth. Pine trees are often classified according to whether they have pine needles, which come in clusters of two, three, or five.

Phyllotaxis is a property, which is found, commonly, not only in leaf arrangement on a stem, and in the arrangement of seeds in a flower head, or the number of petals in a flower, but, also, in the number of filaments and stigmas a flower head has, the number of sepals a flower head has, the number of leaves at any given point on the stem of a plant (e.g., 1, 2, 3, 4 (or 2*2), 5, etc.), and the total number of veins on either side of the center rib of the leaf of a plant. For instance, a dandelion flower will often have thirteen unfolded sepals at the base of its flower head, consisting of a top layer of six unfolded sepals, and a bottom layer of seven unfolded sepals. Many broad-leafed plants have leaves with a number of veins on either side of the center rib of the leaf, being 1, 2, 3, 5, 8, 13, or 21.

Almost all types of plants seem to display some type of phyllotaxis based on the Fibonacci numbers. Surely, upon these observations, it can be stated, as one part of a plant is merely an extension, or replica of another part of the plant, that the Fibonacci numbers, as a key to the growth patterns, or geometric design of most plants, play a fundamental role in the mathematics of the genes of plants.

The role of the Fibonacci numbers in phyllotaxis, or the arrangement of leaves on a stem, also, allows for a maximum exposure of the surface of each leaf to sunlight, and, therefore, is the most energy efficient. This is true, especially, as leaves in plants, and other plants overlay leaves. According to C. W. Wardlaw, broad-leafed plants have a low phyllotaxis ratio, allowing plenty of room between consecutive leaves, and greater exposure to sunlight. Narrow-leafed plants have a higher phyllotaxis ratio, allowing less room between consecutive leaves, and less exposure to sunlight.

Generally, phyllotaxis makes a plant stronger and more durable, allowing for its continued survival and evolution. I am indebted to Thompson (1942) for the following account of Leonardo da Vinci's interpretation of the phenomenon of phyllotaxy: "Leonardo had in like manner explained the leaf-arrangement as serving to let air pass between the leaves, keep one from overshadowing another, and let rain-drops fall from one leaf to the one below." (p. 932)

I am, also, indebted to Thompson (1942) for the following account of how phyllotaxy was, perhaps, one of the first studies of science of man, including the Egyptians:

> The beautiful configurations produced by the orderly arrangement of leaves or florets on a stem have long been an object of admiration and curiosity; and not the least curious feature of the case is the limited, even the small number of possible arrangements which we observe and recognise. Leonardo da Vinci would seem, as Sir Theodore Cook tells us, to have been the first to record his thoughts upon this subject; but the old Greek and Egyptian geometers are not likely to have left unstudied or unobserved the

spiral traces of the leaves upon a palm-stem, or the spiral order
of the petals of a lotus or the florets in a sunflower. For so, as old
Nehemiah Grew says, "from the contemplation of Plants, men
might first be invited to Mathematical Enquirys*." (p. 912)

Again, we notice that 5, 8, and 13 are the central Fibonacci numbers in
phyllotaxis (e.g., the pineapple, the pine cone, and the flower heads of certain
plants), perhaps the central numbers of the numerology of the Major Arcana
of the Tarot cards, and the numerological key behind the altitude of the Great
Pyramid of Cheops.

All plants contain a great deal of genetic variation, just as all human beings
have different finger prints. Also, all plants grow under different weather,
and in different soil conditions. Perhaps, these are some reasons why there is
variation between plants of the same species. Evolution has bred plants to be
the most structurally fit, and not, necessarily, for some phyllotaxis show, based
on some precise and clear cut Fibonacci ratios. Although this is the case for
flowers, it is, generally, not the case for the general body of the plant, which,
usually, has a much more complex and abstract structure.

I

The Historical Development of the Golden Section from Ancient Egypt

C. The Great Pyramid of Cheops

1. Introduction

The Great Pyramid of Cheops is an ancient megalithic monument, built around 2500 B.C. It has four equivalent, isosceles triangular sides, and an, approximately, square base. It is called a pyramid, because it has a pyramid shape. It, originally, had flat, planed sides, covered with a fine mantle of limestone, before it was defaced by the Arabians, during the thirteenth and fourteenth centuries, A.D. According to one Arab historian, Abd-al-Latif, who lived during the thirteenth century A.D., the limestone mantle of the Great Pyramid was, originally, completely, inscribed with Egyptian hieroglyphics.

The Great Pyramid of Cheops was the biggest of the Egyptian pyramids ever built. It is the biggest of the three pyramids erected on the Giza plateau. It was, also, the first built there. It rests on the Giza plateau on the west banks of the Nile, approximately, ten miles west of the city of Cairo.

It is called the Great Pyramid of Cheops, because it is thought to have been built during the reign of Pharaoh Khufu. Cheops was an early Greek name for this ruler. The other two pyramids on the Giza plateau are thought to have been built by his son and grandson, Kephren and Mykerinos, respectively. These kings ruled, during the fourth dynasty (c. 2613-c. 2494 B.C.). There were eight kings in the fourth dynasty.

The length of its base is, approximately, 1.57 . . . times the height of its altitude. The average ratio of base to height of all the pyramids on the Giza plateau is eight to five. 8:5 is the ratio of two consecutive Fibonacci numbers, and equals 1.6, which is a very close approximation of the golden ratio. 8:5 was a ratio, commonly, used by the Egyptians in their art, architecture, and furniture design, as was 5:3.

Supposedly, it took twenty years to build. Over 50,000 Egyptian farmers were drafted each year, during the summers when the Nile was flooded, and their crops were not growing. They were organized into gangs to transport the stone from the quarries to the building site. Many of the Egyptian farmers were killed each year. There were other classes of workers, as well, including stone cutters, masons, surveyors, mortar makers, carpenters, and general laborers. They worked year around.

The Great Pyramid rests on land, which stretches further east, west, north, and south than any other point on earth. It rests just, slightly, south of the 30° north latitude meridian. The four faces of the Great Pyramid were oriented to lie directly east, west, north, and south. The descending and ascending passages of the Great Pyramid lie at an angle of 26° 17' from the horizontal, which is, exactly, the angle from the horizon from which the north star can be seen every night. In fact, the descending passage was designed so that an observer, standing at the bottom, could see the north star through the open doorway at the end of the tunnel at night.

The Great Pyramid of Cheops consists of 201 layers of limestone blocks, each weighing, approximately, five tons. There is more limestone and granite in the Great Pyramid of Cheops than there is in all the Christian churches, cathedrals, and chapels built in England, since the time of Christ. It was when still complete, very, nearly, 484' 5", or 5,813" tall.

Limestone has a wide variety of uses and applications. Limestone is soft, and can be easily carved in. The hardest limestone forms marble. The softest limestone forms chalk.

During the time of Napoleon, French savants visited Egypt with Napoleon's troops. Among other things, they discovered the Rosetta Stone in Rosetta, Egypt, which was, later, transported to the British Museum in London. One of these savants was Edmé-François Jomard, who determined that the exact length of the apothem of the Great Pyramid was six hundred feet, or one stadium, an ancient, classical unit of measure. The apothem is the line running from the apex to the midpoint of one of the base sides of the Great Pyramid. Also, he determined that the length of the perimeter of the Great Pyramid was equal to, exactly, half a minute of longitude. He was the first of the European explorers to make any significant discoveries about the dimensions of the Great Pyramid.

The Great Pyramid of Cheops is the first wonder of the Seven Wonders of the ancient world. It was used as a religious temple; a theodolite, or object used for surveying; a geodetic marker; a giant sundial and calendric predictor; an astronomical observatory, during its construction; and last of all, but not least, a burial grounds for the pharaoh for whom it was built, and the treasures of his afterlife. It was, also, used as a repository for a system of weights and measures, which, because they were adequately preserved, had survived, and are still in use in the western world, today.

I

The Historical Development of the Golden Section from Ancient Egypt

C. The Great Pyramid of Cheops

2. Scientific Discoveries

Coincidentally, the exact length of a year is encoded into the perimeter of the Great Pyramid. The ancient Egyptians, obviously, used a system of measurement based on various measurements of the earth's dimensions, including the circumference of the earth, and the exact length of the polar axis of the earth.

John Taylor was born in London, England in 1781. He was the son of a London bookseller. He was a poet, and essayist, whose regular job was that of editor for the London Observer. In 1854, John Taylor published a book called *The Great Pyramid: Why Was It Built & Who Built It?* For thirty years after the return of Colonel Richard William Howard-Vyse from Egypt, John Taylor examined the results of the measurements of the Great Pyramid, including those of Howard-Vyse, and the French savants. Howard-Vyse was an English explorer, who first visited Egypt in 1836. He excavated the five chambers between the thick granite slabs above the King's Chamber. He cleared the rubble from part of the base of the Great Pyramid, and measured the slope of two of the remaining casing stones. The French savants were men of knowledge, who visited Egypt with Napoleon's troops in 1798.

Taylor found that the perimeter of the Great Pyramid divided by twice its height yielded the value 3.144. 3.144 is very close to π, as π is, to the nearest five decimal places, 3.14159. Taylor wondered if the value of π had, actually, been encoded into the proportions of the Great Pyramid, and, if so, why this had been done. Taylor hypothesized that the Great Pyramid represented the Northern hemisphere of the earth: the height representing one-half of the polar axis of the earth, and the perimeter representing the circumference. Taylor reduced Howard-Vyse's measurements of the perimeter to altitude to 366 to 116.5, noticing that 366 is a very close approximation to the exact number of days in a year. Taylor, then recognized that the exact measure of the perimeter of the Great Pyramid, closely, approximated 100 times 366 British inches. That is, one side of the Great Pyramid, closely, approximated the measure of 25 times 366 British inches.

Sir John Herschel (1792-1871) proposed an inch one-thousandth of an inch longer than the standard British inch, based on the measurement of the polar axis of the earth. Herschel found that 500,500,000 British inches fit into the polar axis of the earth, while 500,000,000 would be a better measure. I am indebted to Tompkins (1971) for the following account:

> According to Herschel the only reliable basis for a standard measure was the polar axis of the earth—the straight line from pole to pole—which a recent British Ordnance Survey had fixed at 7898.78 miles (by taking the mean of all the available meridians measured). This translated into 500,500,000 British inches, or an even five hundred million inches if the British inch were half a human hair's breadth longer. (p. 73)

If the polar axis of the earth measured 500,000,000 inches, then half of this would be 250,000,000 inches. Herschel deduced that fifty inches would make a suitable yard, and twenty-five inches would make a suitable cubit. Then, ten million cubits would fit into half the polar axis of the earth. It is known that the Egyptians used several cubits of different lengths for different purposes, including one of 20.63 British inches, and one of 25.025 British inches. These two cubits were first discovered by Sir Isaac Newton. The cubit of 20.63 British inches was discovered from the dimensions of the King's Chamber, which he found to be 20*10 of these cubits. He termed this cubit the "profane" or Memphis cubit. The cubit of 25.025 British inches was discovered from the circumference of the pillars of the Temple at Jerusalem, as reported by the Jewish historian Josephus. He termed this cubit the "sacred" cubit. The length of 25.025 British inches is found elsewhere in the dimensions of the Great Pyramid.

Taylor, also, found that the British Ordnance made their biggest maps on a scale of 1:2,500, which had no relation to the English mile. He found that the side of one British acre measured 100 cubits of 25 inches, each. Also, Taylor found the volume of the sarcophagus found in the King's Chamber measured, almost, exactly, four times what the British farmer still uses for a unit of measure, eight bushels, or the "quarter".

Taylor's calculation for the height of the Great Pyramid of Cheops was 5,813 Pyramid inches. If 5,813 is the number taken for the radius of a circle, then the measure of its circumference will be

$$5,813 * 2\pi = 36,524.156$$

Pyramid inches. This figure divided by 100 yields 365.24156, a close approximation to the exact number of days in a year. Remarkably, 5,813 contains the numbers 5, 8, and 13, three consecutive numbers in the Fibonacci number sequence, and, perhaps, the central Fibonacci numbers in the numerology of the Major Arcana of the Tarot cards. See Figure 3.

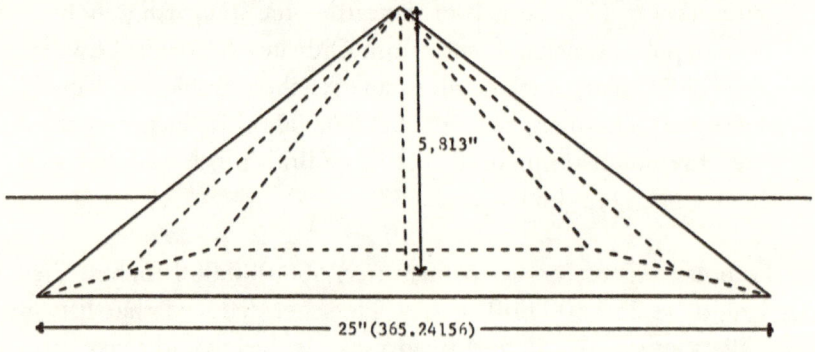

Figure 3

The solar year is the time it takes for the earth to complete one revolution in its orbit around the sun. The length of a solar year is 365.2242 days. The sidereal year is the time it takes a star to return in the same place in the sky, exactly, one year later. The length of a sidereal year is 365.25636 days. So, 5,813 times 2π, divided by 100, falls, somewhere, between 365.2242 and 365.25636, or between the length of a solar year, and the length of a sidereal year, which is, only, 20 minutes later. David Davidson proposed that the

Egyptians included the lengths of the solar year, the sidereal year, and the anomalistic year in the perimeter of the Great Pyramid.

When Howard-Vyse was excavating the chambers between the five layers of huge granite blocks, that lie above the King's Chambers, he found some hieroglyphics indicating the number 17. This meant that the Pyramid had reached "that stage in the seventeenth year of the King's reign" (Tompkins, 1971, p. 64). This is a coincidence, because in the Major Arcana of the Tarot cards, the seventeenth card is the Star card. Seventeen is the seventh prime number, and the sum of ten and seven. The Star card lies before the Moon card, and the Sun card.

Richard Anthony Proctor proposed that the Great Pyramid had been used as an astronomical observatory before its completion. Above the juncture of the passage leading to the Queen's Chamber in the Ascending Passage, the Ascending Passage turns into the "Grand Gallery", which leads to the King's Chamber. It rises 28 feet in seven corbeled layers to a flat roof with removable slabs. Such a "Grand Gallery" would have been ideal for observing, and marking the passage of the stars, planets, and moon across the heavens at night, and the sun, during the day. Proctor, also, proposed that the notches, placed at regular intervals on the floor along either side of the "Grand Gallery", could have been used to erect scaffolding and benches where observers could, easily, sit.

Also, Howard-Vyse discovered two of the original limestone casing stones at the base of the Great Pyramid. Each had a slope of 51° 51', ending the mystery of the exact slope of the faces of the Great Pyramid, and adding further information as to the dimensions of the Great Pyramid. The dimensions of each block were 12' by 5' by 8'. This is remarkable, because 12 is the exact number of lunar cycles in one year, and 5 and 8 are two consecutive Fibonacci numbers. The ratio of 8 to 5 is equal to 1.6, which is a close approximation of the golden ratio.

I

The Historical Development of the Golden Section from Ancient Egypt

C. The Great Pyramid of Cheops

3. The Meridian Triangle

If Taylor's theory were true, then the ratio of the length of one base side to altitude would be $\pi/2$, or a close approximation of the golden ratio (see Figure 4). This is one of the reasons why the dimensions of the Great Pyramid of Cheops seem so perfect, and why it must have been so beautiful when it was first built. The Great Pyramid of Cheops, it seems, is very much a man made mountain, used to enshrine and encode the mathematical secrets of the ancient Egyptians. Being the discoverers of these mathematical formulas, the ancient Egyptians were anxious to enshrine them, and to preserve them for future generations to come for both Egyptians, and people of foreign countries. As the Egyptians worshiped the sun, they desired to hold up to God the mathematical perfection, which they thought reflected Him.

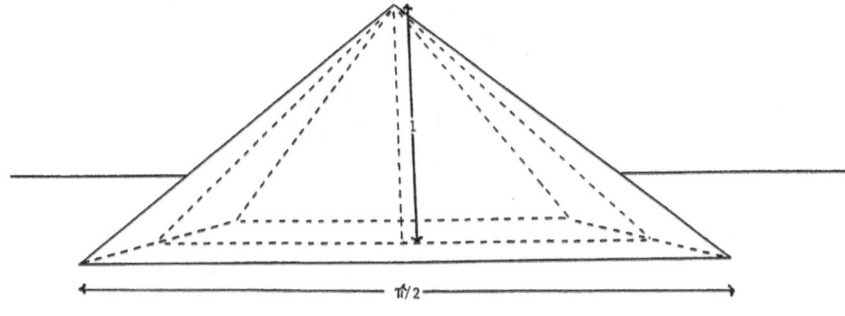

Figure 4

A meridian circle is a circle, which circumscribes the earth and passes through both the North and South poles. Since the Great Pyramid was oriented to lie, directly, north and south, the isosceles triangle, formed by the two apothems on the north and south faces, and the line running between the midpoints of the bases of the north and south faces, is called the meridian triangle. Half of this triangle is a right triangle (see Figure 4), and its proportions are a result of the working of the Pythagorean theorem. It must have been know to the ancient Egyptians, as early as the planning of the Great Pyramid of Cheops.

Its proportions can easily be determined from Taylor's formula. If the ratio of the length of one base side to altitude is $\pi/2$, then the ratio of the length of one-half of one base side to altitude is $\pi/4$. This, immediately, forms the ratio of the lengths of two of the sides of our right triangle, in fact, the legs. The relative length of the hypotenuse of this right triangle can be found using the Pythagorean theorem. Call the length of the altitude, a, the length of one-half of one base side, b, and the length of the apothem, c, (see Figure 5) then

$$a^2 + b^2 = c^2, \text{ or}$$
$$1^2 + (\pi/4)^2 = c^2, \text{ or}$$
$$c = \pm\sqrt{(1+(\pi/4)^2)}$$
$$= \pm\sqrt{(1+.6168501\ldots)}$$
$$= +1.2715542\ldots$$

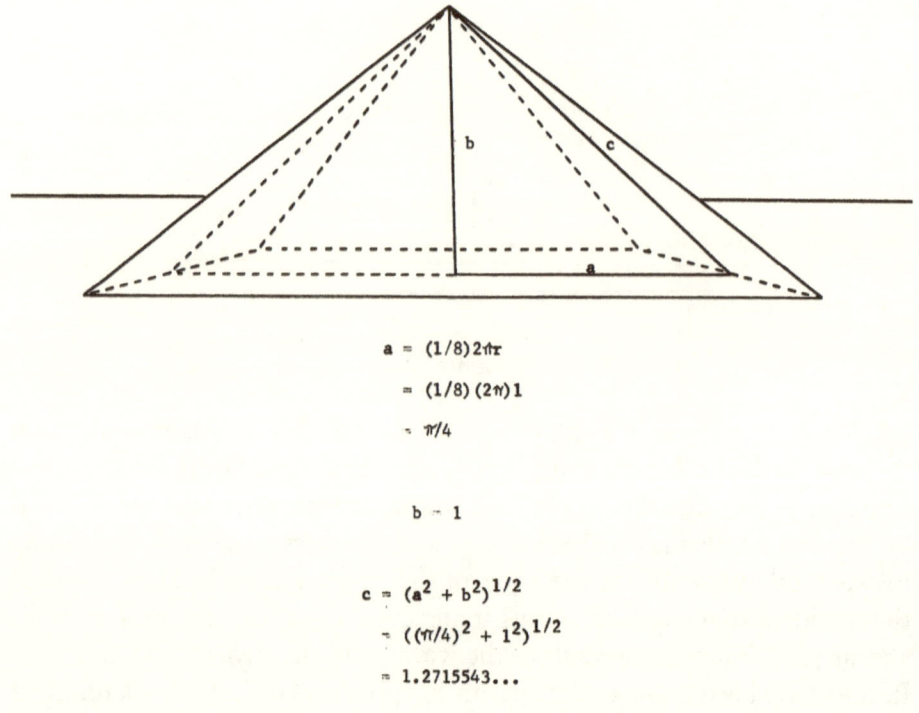

$$a = (1/8)2\pi r$$
$$= (1/8)(2\pi)1$$
$$= \pi/4$$

$$b = 1$$

$$c = (a^2 + b^2)^{1/2}$$
$$= ((\pi/4)^2 + 1^2)^{1/2}$$
$$= 1.2715543\ldots$$

Figure 5

Therefore, the ratio of altitude to the length of one-half of one base side is equal to $1/(\pi/4)$, or $4/\pi$, or $1.2732395\ldots$ The value of the square root of the golden ratio is $1.2720196\ldots$ Therefore, this value is a very close approximation of the value of the square root of the golden ratio. Also, the ratio of the length of one apothem to the altitude is equal to $1.2715542\ldots$ /1, or just $1.2715542\ldots$, again, a very close approximation of the value of the square root of the golden ratio. The average of these two ratios is $(1.2732395\ldots + 1.2715542\ldots)/2$, or $1.2723968\ldots$

Also, with a meridian triangle of these proportions, the measure of the angle of the slope produced is 51° 51'. This is, exactly, what Howard-Vyse measured for the casing blocks at the base of the Great Pyramid, 51° 51'.

The length of one base side of the Great Pyramid, as measured by William Flinders Petrie, is 9069 inches. With a slope of 51° 51', as measured by Howard-Vyse, this would create a height of 5773.5 inches. The relative length of one royal cubit, as found by Sir Isaac Newton, is 20.63 inches.

In 1883, Petrie published a book called *The Pyramids and Temples of Gizeh*, summing up his finds. He proposed that the base of the Great Pyramid was designed to be, exactly, 440 royal cubits, and the altitude of the Great Pyramid was designed to be, exactly, 280 royal cubits. These dimensions can be, quickly, reduced to a ratio of 110 to 70 (see Figure 6). Therefore, if the relative length of one-half of one base side is 55, the relative height is 70. Using these proportions for the right triangle formed by one-half of the meridian triangle, the relative length of the apothem can, easily, be found. Call the length of the altitude, a, the length of one-half of one base side, b, and the length of the apothem, c, (see Figure 6) then

$$a^2 + b^2 = c^2$$
$$70^2 + 55^2 = c^2$$
$$\therefore \quad c = \pm\sqrt{(70^2 + 55^2)}$$
$$= \pm\sqrt{(4900+3025)}$$
$$= +89.022469 \ldots$$
$$\cong 89$$

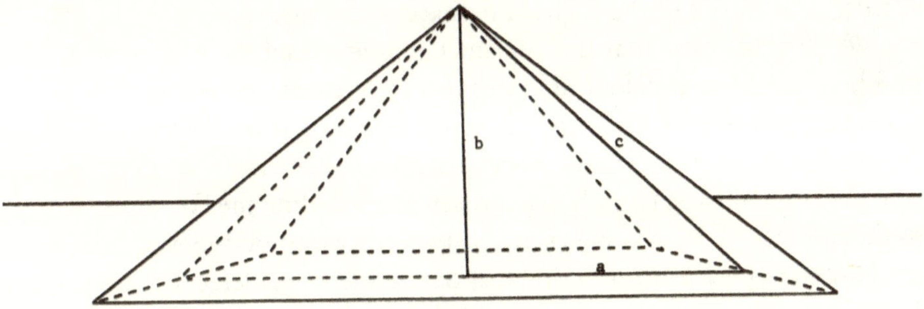

r (the royal ell) = 20.63"

a = 55 x 4r = 220r

= 4,538.6"

b = 70 x 4r = 280r

= 5,776.4"

c = 89 x 4r = 356r

= 7,344.28"

Figure 6

55, 70, 89 is a Pythagorean triple with fairly large numbers involved. It is the Pythagorean triple with the largest numbers involved, that western scientists know, the Egyptians used. 55 and 89 are two consecutive Fibonacci numbers. It is well known that one of the properties of the Fibonacci number sequence is that the sum of the squares of two consecutive Fibonacci numbers is equal to another Fibonacci number. Perhaps, the Egyptians had this property of the Fibonacci numbers in mind when they designed the dimensions of the Great Pyramid in terms of cubits.

Further evidence exists to give credence to this theory. We already know that, according to Taylor's theory, the altitude of the Great Pyramid was designed to be 5,813 Pyramid inches tall. 5, 8, and 13, again, are three consecutive Fibonacci numbers, and their combination in such a manner already gives us the dimension of one of the sides of this right triangle.

The Fibonacci numbers were, commonly, used by the ancient Egyptians in the proportions of the designs of much of the art and architecture, including in the design of tomb furniture, as a rough approximation of the golden ratio. But, in these cases, the lower Fibonacci numbers were used, especially, 8 to 5, and 5 to 3. With a building as grand as the Great Pyramid, surely, the ancient Egyptians were interested in using, and preserving their knowledge of the mathematics of the higher Fibonacci numbers.

We already know that the ancient Egyptians used the higher Fibonacci numbers in the numerological key of the Tarot cards, including twenty-one (the exact number of cards in the Major Arcana), and fifty-five (the exact number of cards in the Minor Arcana, minus one, or the Fool card). Again, seventy-seven, the total number of cards in all, excluding the Fool card, is equal to the product of seven and eleven, the two lucky numbers. Seven and eleven are two consecutive prime numbers. Seven was considered by the ancient Egyptians to be the virgin number. It being the exact number of moving heavenly bodies visible to the naked eye at night. Eleven was considered to be the absolute number. 55 and 89 are Fibonacci numbers, commonly, found in the larger flower heads of sunflowers.

Including the Fool card, there are seventy-eight cards in both Arcanas. Seventy-eight is equal to the product of six and thirteen, or two, three, and thirteen. Thirteen is the sixth prime number. In the Major Arcana, the thirteenth card is the Death card. To many, thirteen is an unlucky number (Friday, the thirteenth).

Although 70 is not, exactly, equal to the square root of (89^2-55^2), it is very close. As for how the early Egyptian priests and/or architects had, originally, planned the design for its shape, we can be sure they kept this Pythagorean triple in mind. Also, with a right triangle forming half of the meridian triangle of the Great Pyramid of these dimensions, the ratios of the dimensions of adjacent sides yields values close to the golden ratio, and the square root of the golden ratio. That is, the ratio of the altitude to the length of one-half of the base side is equal to 70/55, or 1.27 . . . Again, as with Taylor's theory, this ratio is very close to the value of the square root of the golden ratio, 1.2720196 . . . The ratio of the length of one apothem to the length of one-half of one base side is equal to 89/55, or 1.618 . . . , a very close approximation of the golden ratio. The ratio of the length of one apothem to the altitude is 89/70, or 1.2714285 . . . , again, a very close approximation to the value of the square root of the golden ratio. The average of these two ratios, $((70^2 + 55^2)^{1/2} + (89^2 + 70^2)^{1/2})/2$ is equal to 1.2720778 . . . , which is, almost, exactly, the value of the square root of the golden ratio.

Also, with a meridian triangle of these proportions, the measure of the angle of the slope produced is 51° 51'. This is, exactly, what Howard-Vyse measured for the casing blocks at the base of the Great Pyramid, 51° 51'.

And, thus, through the Fibonacci numbers, the proportion of the golden ratio plays a very important role in the dimensions of the Great Pyramid, especially, to an observer standing on the ground, who can, directly, see the relationship between height and the length of one base side, or the meridian triangle. Through the formula for the circumference of a circle, the ancient Egyptians have found, and presented to man for all time, a close relationship between the value of π, and the value of the golden ratio in the meridian triangle of the Great Pyramid of Cheops.

This right triangle represents the Egyptian knowledge of the Pythagorean theorem, and of Pythagorean triples, which were used, commonly, elsewhere in Egyptian architecture. The Egyptians knew of several Pythagorean triples. One of these was the 3, 4, 5 Pythagorean triple, which was, commonly, used by the Egyptians in laying out plots of land. That is, they would tie a rope with twelve knots in it, and make it into the form of a right triangle with the sides of the respective lengths, 3, 4, and 5.

I

The Historical Development of the Golden Section from Ancient Egypt

D. Pythagoras

1. A Bibliography

Pythagoras was a Greek mathematician, philosopher, and political leader. He was born on the island of Samos around 580 B.C. Samos was a Greek island, and it is located just off the west coast of Turkey. He is believed to have been a student of Thales and Anaximander. He is, sometimes, known as the father of Greek mathematics. Pythagoras is reputed to have traveled wide, including to Egypt, and, perhaps, the East. According to one account, given by Tons Brunés, Pythagoras had lived for twenty-two years in Egypt.

> Though Greece has been looked upon as the birthplace of mathematics largely because of surviving written material on the subject of mathematics and geometry—Brunes points out that Pythagoras, the founder of Greek mathematics, spent 22 years in Egypt as a priest of the temple (the Great Pyramid), and only returned to Greece after Cyrus the Great, king of Persia, burnt the temples at Memphis and Thebes in 527 B.C. and dragged him off as a prisoner to Babylon.
> Back in Greece, Pythagoras taught mathematics on the basis of what he had learnt in Egypt; but after his death his followers

were persecuted and had to take refuge abroad. Some eighty years later, Plato left Athens after the execution of Socrates and joined the Pythagorean societies. He traveled to Egypt, where he too was initiated into the lower degrees of learning in the temple, which were slowly recovering from being disbanded by the Babylonian-Persian conquerors. (Tompkins, 1971, p. 262)

Pythagoras had established a brotherhood in Calabria, which is the land forming the southwest part of the boot of Italy, and Sicily. This general land area was known as Magna Graecia. This brotherhood was an ascetic group, which believed in the betterment of mankind through learning in mathematics, philosophy, and astronomy. They believed that animals could speak, and that numbers formed the basis of the universe. They, even, shared all their material possessions in common.

Within the Pythagorean brotherhood, there were three stages of initiation. The first stage consisted of the "Politics", who were early initiates in the brotherhood. The second stage consisted of the "Nomothets", who were the philosophers in the group, and who directed the social and political activities of the brotherhood. The third stage consisted of the "Mathematicians", who had completed their training in initiation, including the "Laws of Number".

But, Pythagorean teaching had become so popular in Magna Graecia, that the Pythagoreans had soon entered, and taken over politics in this area. They were a group of people, who desired to support the aristocrat party in Italy. But, the democrats, soon, took over, and forced harsh rule over Italy. Pythagoras was forced to leave Crotona by Cylon, who started an uprising against the Pythagoreans. Pythagoras moved to the Greek colony at Metapontum on the southern coasts of Italy, where he spent the last ten years of his life. Pythagoras died in 497 B.C.

The Pythagorean brotherhood had retained political supremacy in Magna Graecia, until about 450 B.C. At this time, many of the vassal cities had become detached from the Crotoniate League, and a giant fire in Metapontum killed many of the current leaders of the Pythagorean school. By 350 B.C., the Pythagorean brotherhood had all but disappeared.

Pythagoras had devoted much of his lifetime to the study, and pursuit of new ideas in mathematics. For example, the fourth book of Euclid is devoted, exclusively, to the study of Pythagorean mathematics. Pythagoras was the first Greek mathematician to propose the concept of irrational numbers. He found that in a square with sides of length one, the length of the diagonal would be equal to the square root of two. He used the Pythagorean theorem.

Pythagoras was, also, the first Greek philosopher to propose that the world was round.

If Pythagoras lived during the sixth century B.C., and if the Greeks adopted the ideas of the golden section from the Egyptians, "some time during the sixth century B.C." (Hambidge, 1919, p. 2), then why could not have Pythagoras obtained his ideas from the Egyptians? These include the relationship between the pentagram and the golden section, the Pythagorean theorem, and a musical scale, based on the Fibonacci numbers, or the golden section. Obviously, if the Greeks were obtaining knowledge from the Egyptians at this time, the time would have been ripe for Pythagoras, and the Pythagoreans to acquire knowledge of mathematics from the Egyptians.

> Jomard pointed out that Herodotus, Plato, Diodorus and many others had all named Egypt as the birthplace of geometry, that Solon as well as Plato had come to Egypt to study geometry, and that Pythagoras had learnt from the Egyptians his theorems of geometry, his art of calculation, and his doctrine of metempsychosis. (Tompkins, 1971, p. 48)

I

The Historical Development of the Golden Section from Ancient Egypt

D. Pythagoras

2. The Pythagorean Scale

The Pythagorean scale is based on the ratio of 2/3 to ¾, or 8/9. That is, ¾ times 8/9 is equal to 24/36, or 2/3. If one divides a scale into ½ and 1, then the point half way between these two points is ¾ (see Figure 7). Now, if one divides this scale further at a point equal to 2/3, the scale will be divided into three separate parts. The object of this scale is to divide the total length given into eight separate lengths. This can be done by dividing the total length of the scale into two separate tetrachords, or series of notes, consisting of four notes. This was a common practice carried out before Pythagoras' time, but he was the first to do so in such a manner.

Call the first note of length one, C, and the last note of length one-half, C octave, then we can build the C scale, or the major scale of the Pythagorean scale. Next, multiply the string of length one by 8/9, which is our primary ratio. Of course, this will produce a string of length 8/9. Call it D (see Figure 7). Next, multiply this length, again, by 8/9. This will produce a string of length 64/81. Call it E. Now, the difference between this length and ¾, which is the next lower note, is 243/256. This difference is less than the difference produced between note E, and a length produced by multiplying the length of note E by 8/9. This is why in the Pythagorean scale, we have two long notes followed by one short note. Call the note of length ¾, F. Now, we have built the first tetrachord.

We already know that ¾ times 8/9 is equal to 2/3, so we need no extra notes between these two notes. Call the note of length 2/3, G (see Figure 7). To build the second tetrachord, we follow the same procedures used to build the first tetrachord. Multiply the length of note G, 2/3, times 8/9. This produces a string of length 16/27. Call it A. Now, multiply the length of this note by 8/9. This produces a string of length 128/243. Call it B. Again, as with the first tetrachord, the difference between the second to last, and the last notes, or between 128/243 and ½ is 243/256. Again, we have two long notes followed by one short note.

Now, we have built a scale, or octave consisting of seven notes. Call it the C scale. An octave can, as it often did in the past, consist of 5, 6, 7, 8, or 9 separate notes. Three is the number of notes commonly played in one chord. Five is the least number of notes for one octave. And, eight is the number of notes contained in a double tetrachord, or common subdivision of an octave. 3, 5, and 8 are three consecutive Fibonacci numbers. Also, 7, the number of unique notes in one octave of the Pythagorean scale is the virgin number. Also, 7 is the average between 5 and 9, the least and greatest number of notes possible in one octave, respectively.

Also, 1, ½, and 2/3 are the ratios of the smaller to the larger of two consecutive Fibonacci numbers: 1:1, 1:2, and 2:3, respectively. Each approaches a closer approximation to the value of the golden ratio minus one. 2/3 is $.\overline{6}$. The value of the A note, 16/27, is .59259, which is an even closer approximation to the value of the golden ratio minus one. It forms an even closer approximation to the value of the average of $\pi/2$ and the golden ratio, minus one, $((\tau+\pi/2)/2) -1$, or .5944151 . . . These are combinations of Fibonacci numbers, that are found, commonly, in nature, or phyllotaxis, the arrangement of leaves on a stem.

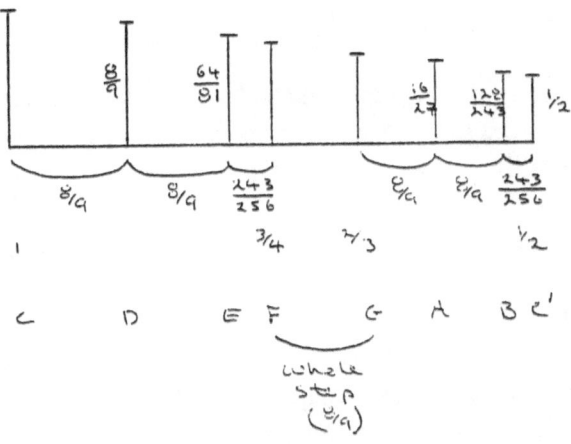

Figure 7

Notice that ¾ and 2/3 are proportions, commonly, used in art. 2/3 divides the length of one unit into three separate parts. ¾ divides the length of one unit into four separate parts. The difference between ¾ and 2/3 is .083. The difference between 2/3 and the value of the golden ratio minus one is .04863269 . . . The difference between the value of the golden ratio minus one and the value of π/2 minus one is .04723765 . . . This produces an interesting correlation between ¾ and 2/3, the value of the golden ratio minus one, and the value of π/2 minus one, which any artist should take note of.

Again, we return to the relationship between 1 2/3, the golden ratio, and π/2. In this series, we find that the average of 1 2/3 and π/2 is 1.6187315 . . . , which is a very close approximation of the golden ratio (see Chapter I., Section F.). π/2 is a proportion, which is found, commonly, in the circling of the square series.

Now, from the C scale, we can create other scales, including the G, D, A, E, B, and F scales. For example, in order to create the G scale, we can multiply each note in the C scale by 2/3, which is the value of G in the C scale. Then, we get a scale with notes of the following lengths: G, 2/3; A, 16/27; B, 128/243; C', ½; D' (or D octave), 4/9; E', 32/81; (F#)', 256/729; and G', 1/3. The notes G, A, B, and C' in the G scale correspond to G, A, B, and C' in the C scale. Now, D' (or D octave) is a length equal to, exactly, one-half the length of D. E' is a length equal to, exactly, one-half the length of E. But, (F#)', 256/729, does not produce a length equal to, exactly, one-half the length of F, or 3/8. It is, slightly, less. Therefore, we arrive at our first sharp. It is a shorter note on the Pythagorean scale than F', or, slightly, less than an octave less than F.

We can go on producing scales in this fashion. With each successive scale, we produce one more sharp. For instance, the scale of D has two sharps, and the scale of A has three sharps. The scale of E has four sharps, and the scale of B has five sharps.

With the Pythagorean scale, a C scale has been created, which is, perfectly, in tune. But, when other scales are created from this scale, they are slightly out of tune. Therefore, a method of tempering has been introduced, which, slightly, changes the lengths of some notes in order to accommodate the tonal differences between scales. The changing of the lengths of certain notes in our modern day musical scale brings the notes of the Pythagorean scale closer to the notes of a, strictly, harmonic scale. This harmonic scale is based on the ratios of the successive fractions 1, ½, 1/3, ¼, 1/5, 1/6, 1/7, 1/8, 1/9 . . . The means of basing the lengths of the various notes on a harmonic scale was, also, in common use before the time of Pythagoras.

The method of subdivision of the C scale of the Pythagorean scale is called the Dorian mode, because of its place of origin in the Dorian colony of Crotona on the southern coast of Italy. It is not sure which mode Pythagoras prescribed, but it is sure, that he prescribed that all modes be divided at points of one, three-fourths, two-thirds, and one-half.

Eight-ninths, the major constant of subdivision in the Pythagorean scale, is equal to $2(4/9)$, or $2(2^2/3^2)$, or $2(2/3)^2$. Again, this ratio is based on two-thirds, the ratio of the smaller to the larger of two consecutive Fibonacci numbers, and equal to 1 2/3 minus one.

I

The Historical Development of the Golden Section from Ancient Egypt

E. The Tarot Cards

1. Introduction

The history of the Tarot cards is long, yet, misunderstood. There are many theories about the origin of the Tarot cards. One of these is that the Tarot cards came from Egypt. There are many symbols, and signs, which point to this possibility. First of all, the Hebrew alphabet contains 22 letters, each of which can be associated with one of the 22 cards of the Major Arcana. The Hebrews, of course, were located close to Egypt.

The theory of fortune-telling by using the Tarot rests on a fundamental assumption of magic: that all the phenomena of the universe are connected together in a great design or pattern. The basic theory or structure is basically Greek in origin (particularly Pythagoras) but its essential principals—that all things can be expressed in numbers and that the universe is constructed to a mathematical pattern and this pattern involves opposites and their reconciliation—have received powerful support and their most important magical development in the Jewish system of magic: the Cabala.

The Cabala is a body of occult doctrine which has been enthusiastically used by non-Jewish occultists since the fifteenth century. It consists of numerous writings by various anonymous authors, and for our purposes, the most important work in the Cabala is the *Sepher Yetzirah* (Book of Formation), which was written in Hebrew, probably in Babylonia between the third and sixth centuries A.D.

In the *Sepher Yetzirah*, there is an account of the Creation, and it says: "In the thirty-two wonderful Paths of Wisdom did Jah, Jehovah Sabaoth, the God of Israel, the Elohim of the living, the King of the ages, the merciful and gracious God, the exalted One, the Dweller in eternity, most high and holy—engrave his name by three Sepharim (means of expression)—Numbers, Letters and Sounds." God used numbers, letters, and sounds in creating, says the *Sepher Yetzirah*, because God is man writ in large and the three most important means of human communication is by counting, writing, and making meaningful noises. The "thirty-two wonderful Paths of Wisdom" are the ten sephiroth, or numbers (in the Tarot, the Minor Arcana: Ace-Ten), and the twenty-two letters of the Hebrew alphabet (the Major Arcana: 0-XXI).

The occultists associate the twenty-two letters of the Hebrew alphabet with the twenty-two works of creation in the Book of Genesis and the twenty-two books of the Old Testament. (The occultists always cheerfully assert that there are twenty-two books in the Old Testament, although the Christian version gives many more and the Jewish version lists twenty-four.) In these works of creation, therefore, is the key to all wisdom, all truth, all knowledge of God and the universe. The cabalists associate them with the Twenty-two Paths, which are the roads that lead from one sephira (level of human perfection) to another. For them, the numbers and the Paths account for all that is in the universe, and they are the soul's way to God and the magician's way to power; the steps in the process of spiritual expansion through which man can extend himself to cover the entire universe and control it.

In this book, the meaning of each "trump" of the Major Arcana (The Fool card, The Magician card, etc.) depends on its design and number, but to some cabalists it depends on its position, or Path, on the Sephiroth Tree (Tree of Life). For the cabalists, the system of correspondence which throws additional light on the meaning of each trump and Path is an extended version of the one in the Sepher Yetzirah, which divides the Hebrew letters into groups of 3, 7, and 12, corresponding to the 3 elements (fire, air, and water—there is no category for earth in this system), the 7 planets (using the Ptolemaic universe, not the Copernican universe), and the 12 signs of the Zodiac. (Laurence, 1972, pp. 14-16)

The Greek alphabet contains twenty-four letters. Twenty-four is twice the number of lunar cycles in one year, or half the number of lunar cycles in a given four year period, or within one leap year cycle. There were four years between consecutive Olympic games in Greece.

> Letters were used instead of numbers, and distinguished by accents, ά—1; β′—2; γ′—3 κ.τ.λ. α,—1,000; β,—2,000 κ.τ.λ. But you need not bother about these.
>
> For reference, the full list is as follows:

1—α	10—ι	100—ρ
2—β	20—κ	200—σ
3—γ	30—λ	300—τ
4—δ	40—μ	400—υ
5—ε	50—ν	500—φ
6—ϛ	60—ξ	600—χ
7—ζ	70—ο	700—ψ
8—η	80—π	800—ω
9—θ	90—ϙ	900—ϡ

> This can easily be worked out if you remember that with the exception of 6, 90 and 900 you just go through the alphabet, but if you try to do arithmetic with them you will realize why the Greeks only studied geometry amongst the mathematical sciences. (Hudson, 1960, p. 114)

Second of all, the Major Arcana of the Tarot cards contains a symbology, which is similar to the symbology contained in the religion and hieroglyphics of ancient Egypt.

> All of these accounts of the Tarot's origin add depth and interest to the cards, and none of them can be denied. A profound study of the Tarot does reveal much of the ancient Hebrew wisdom of the Kabala; many of the symbols are indeed linked with Egyptian mythology; and the Gypsies are widely considered to have possessed, from time immemorial, an uncanny gift for reading the past, the present and the future, by means of the Tarot pack. (Gray, 1969, p. 12)

Court de Gebelin in 1781 writing in Volume I of *Le Monde Primitif* presents a strong argument in favor of the Egyptian origin of Tarot cards. According to Gebelin, the twenty-two Major Arcana cards are an ancient Egyptian book, *The Book of Thoth*, saved from the ruins of the burning Egyptian temples. Thoth was the Egyptian Mercury said to be one of the early kings and the mythical inventor of speech and hieroglyphs or letters with its attendant mysticism. Its basis was an alphabet in which all gods are letters, all letters ideas, all ideas numbers and all numbers perfect signs. Many scholars of the occult recognize in tarot cards the pages of hieroglyphic books, containing the principles of the mystic philosophy of the Egyptians in a series of symbols and emblematic figures. Gebelin believed that the esoteric tarot symbols subsequently were spread throughout Europe by wandering gypsies. (Kaplan, 1972, p. 8)

The Tarot pack consists of 78 cards: 56 contained in four suits (the forerunners of modern playing cards), called THE MINOR ARCANA; and 22 additional picture cards, known as THE MAJOR ARCANA. The Major Arcana are said to be derived from the pages of the oldest book in the world, originated by Hermes Trismegistus, councilor of Osiris, King of Egypt, at a period when hieroglyphic writing, magic, astrology, and other mystic sciences flourished. Some scholars maintain that they were invented by the Chinese; others that they were brought from India by the Gypsies. They are also frequently related to the Kabalistic lore of the Hebrews; and a correspondence is often pointed out between the cards of the Major Arcana and the letters of the Hebrew alphabet. (Gray, 1960, p. 11)

Gebelin also described the allegorical meaning of the four suits: "Apart from the trump, this game is made up of four suits represented by their signs. They are called Swords [*épées, piques,* spades], Cups [*coeurs,* hearts], Batons [*tréfles,* wands, clubs], and Coins [*carreaux, pentacles, deniers,* money, diamonds]. Each of these suits consists of fourteen cards, the cards numbered from 1 to 10 and four face cards which we call the King, Queen, Knight, and his Equerry or Knave."

Gebelin ascribed the four suits to the four classes into which the Egyptians were divided.

The sword represented the sovereigns and all the military nobility.
The stick or the club of Hercules represented agriculture.
The cup represented the sacerdotal rank, clergy or priesthood.
The denier represented commerce of which money is the sign.

Gebelin firmly concluded that the game of tarot, therefore, must have been invented by the Egyptians since it is based on the number seven; it pertains to the division of the Egyptians into four classes; most of its trump are definitely connected with Egypt such as the two chief Hierophants, male and female (equivalent to II The High Priestess and V The Hierophant), Isis or the Dog-Star (XVII The Star), Typhon (XV The Devil), Osiris (VII The Chariot), The House of God (XVI The Tower), The World (XXI The World), the dogs which guard the tropics (XVIII The Moon), etc.; and since the game was entirely allegorical as one might reasonably expect from a study of Egyptian art and philosophy. (Kaplan, 1972, pp. 38 and 40)

Third of all, the word Tarot is thought to have Egyptian origins.

Tarot cards are so old that their beginnings are obscured by the many cultures thought to have used them. One theory of their origin, nevertheless, is that the ancient Egyptians invented the Tarot as a repository of their occult lore, and from there, the Tarot was supposedly brought into Europe by the gypsies. In the syllabary of ancient Egypt, *Tar* means "Path" and *Ro* means "royal"; thus, the transliteration of "Tarot" is "Royal Path (of life)." The Egyptians, moreover, considered the Tarot a hieroglyphic "book," rather than cards. The plates or cards of the Tarot originally were worn as amulets about the necks of the high priests, who were the custodians of sacred knowledge. Taken together, the symbols of the amulets constituted the Egyptian "book of life." (Laurence, 1972, p. 11)

The gypsies, according to Gebelin, were in fact Egyptians who dispersed over Europe and from whom we derive the custom of fortune-telling with cards. Gebelin saw the word tarot as a combination of tar signifying way or road, and ro, ros, or rog, implying king or royal; the word tarot meaning, therefore, the "royal road of life." (Kaplan, 1972, p. 35)

I

The Historical Development of the Golden Section from Ancient Egypt

E. The Tarot Cards

2. The Number Seven

Gebelin has proposed that seven is the whole numerological foundation behind the Tarot cards. I set out to prove, or, rather, reprove this theory.

> According to Gebelin, tarot cards were either an allegory expressed in ancient Egyptian hieroglyphics relating to their philosophy and religion, or a book presenting the history and creation of the world and of the first three ages, commencing with Mercury himself. The four suits were meant to represent the four states or orders of political society. The whole foundation of the tarot cards was resolved from various aspects into the number seven which was sacred to the Egyptians and upon which they based the elements of all sciences. Each suitor color was composed of twice seven cards. The *atouts* number three times seven, and the total number of cards is seventy-seven, The Fool being "0." (Kaplan, 1972, pp. 34-35)

First of all, we know that there are fourteen cards in each suit of the Minor Arcana. Fourteen is twice seven, and is the same number of days in one two week period, or fortnight. There are twenty-four hours in a day, instead of twelve. There are twenty-four months in one two year period, or half of a leap year cycle. Two is the first prime number, and the third Fibonacci number. Also, two is the number of pairs.

Second of all, there are fifty-six cards in the Minor Arcana. Fifty-six is equal to seven times eight. Eight is the sixth Fibonacci number, thus relating it to the number six, or the base six number system.

Third of all, the numbers twenty-one and fifty-six add up to seventy-seven. Seventy-seven is equal to seven times eleven. Seven is the fourth prime number. Also, seven is equal to the sum of two (the High Priestess card), and five (the Hierophant card), two alternate Fibonacci numbers. Eleven is the fifth prime number, thus relating it to the number five. Five is the number of the pentagram. "Five was the number of Love (uniting Two, the first even, female, number, and Three, the first odd, male, number)" (Ghyka, 1977, p. 113) to the Pythagorean mystics. Eleven is, also, equal to the sum of the two Fibonacci numbers, three and eight. Seven and eleven are considered by many to be the two lucky numbers.

Fourth of all, there are twenty-one whole numbered cards in the Major Arcana. Twenty-one is the eighth Fibonacci number. Twenty-one is equal to three times seven. Three is the second prime number, and the fourth Fibonacci number. Also, recall, three is considered the number of masculinity.

To many ancient civilizations, seven has been considered a sacred number. "In Pythagorean Number-Mystic, seven was the Virgin-Number." (Ghyka, 1977, p. 21) Asimov proposes why seven has been considered the sacred number.

> The week is an artificial unit. It may have been meant to mark the coming of the chief phases of the moon: new, first quarter, full, and last quarter. These are 7.38 days apart. To be strictly lunar, the week would have to be sometimes seven days long and sometimes eight. The people of the Tigris-Euphrates Valley, who were the most advanced astronomers in the early days of civilization, were probably influenced, however, by the sun, the moon, Mercury, Venus, Mars, Jupiter, and Saturn, the seven heavenly bodies easily visible to the unaided eye that moved in complicated fashion against the fixed-pattern background of the stars. By granting divine honors to the sun, the moon, and the planets and allowing each one day, the notion of the invariable seven-day week arose. It was passed on to the Jews, then to the Christians, and is still with us. It is a very inconvenient unit that does not fit evenly with either the month or the year. (Asimov, 1982, p. 184)

"The mystery of the seven stars which thou sawest in my right hand, and the seven golden candlesticks. The seven stars are the angels of the seven churches: and the seven candlesticks which thou sawest are the seven churches." (Revelation 1: 20)

Also, notice, the seventh card of the Major Arcana is the Chariot card. Like the Hierophant card, the Chariot card is a key to the Egyptian numerology, and mysticism behind the Tarot cards.

> Gébelin, a friend of Benjamin Franklin, traced the origin of the cards back to ancient Egypt, identifying them with the Book of Thoth, the scribe of the Egyptian gods. His erroneous description and interpretation of the cards even entered the Morgan files, where a note on the triumphal car states that the customary representation was Osiris in his triumphal car, the symbol of war in the age of bronze. (Voelkle, Jan./Feb., 1985, p. 43)

It would make sense that the symbol of the chariot would be an important symbol to the Egyptians, because the wheel was invented by the Sumerians.

> Wheeled vehicles were probably first developed in a broad, roughly, trapezoidal area with its longer base extending from north of the Black Sea to the Caspian and its shorter base the northern end of the Persian Gulf, with Lake Van in eastern Asia Minor as the center. The earliest wheeled vehicles have been found within 600 miles (966 km) of the lake. The oldest archaeological evidence indicates that the wheeled vehicle came into existence somewhat earlier than 3000 B.C. The earliest of these were probably two-wheeled wooden carts built by the Sumerians in the forested regions south of the Caucasus and Tarsus mountains. Four-wheeled vehicles with draft poles recently found north of the Caucasus Mountains in the U.S.S.R. date from about 2400 BC. The wheeled vehicle apparently was taken westward into Europe by people who raveled up the Danube River and north to the Balkans where there is evidence for wheeled vehicles going back beyond 2000 BC. (Benson, 1995, Vol. 26, p. 318)

The wheel did not come into use in Egypt, until about 1600 B.C. When it did first come into use, it must have made a strong impact on Egyptian society, and the royalty, that was in existence at the time. "The Great Pyramid of Giza, built about 2500 BC, before the Egyptians knew the pulley or had wheeled vehicles" (Kranzberg, 1995, Vol. 29, p. 919).

Because the wheel was one of the newest, and most important inventions of the time, it is clear why the image of the chariot would be included as the image of the seventh card of the Major Arcana. The chariot was, popularly, used, at least by the royalty, for sport and hunting. "Tutankhamun in his war-chariot attacking the Syrians. Painted on the stuccoed side of a wooden chest." (MacQuitty, 1976) King Tutankhamun lived during the 14th century B.C.

In the early Tarot card decks from Italy and France, in the Major Arcana, many of the symbols had a different order. They, also, often included a smaller, or greater number of cards. The earliest known deck comes from Gringonneur of France. In the Gringonneur deck, there were seventeen cards in the Major Arcana. They had no numbers, or letters, or inscriptions, so it is not known, exactly, what order they were meant to be placed.

> In the accounts book of Charles Poupart, Treasurer to Charles VI of France, there exists a passage which states that three packs of cards in gilt and variously ornamented were painted by Jacquemin Gringonneur for the amusement of the King of France in the year 1392. Gringonneur was paid 56 sols of Paris. (Kaplan, 1972, pp. 14 and 15)

> However, Waite also transposed, probably without adequate justification, the two cards of Strength and Justice. Strength, generally shown in tarot decks as number XI, is instead designated by Waite as number VIII. Justice, commonly shown as number VIII, is switched by Waite to number XI. (Ibid., pg. 62)

Eight is twice four, or the number of vertices of a cube, and the number of faces of a regular octahedron. Dr. Arthur Edward Waite lived from 1857 to 1942.

The fourteenth card is the Temperance card. Fourteen is twice seven. "Life is hectic in our current age. As a warrior astride his *Chariot* drawn by two horses pulling apart, we rush towards ultimate failure or triumph, often forgetting the cardinal virtues of *Temperance, Justice*, and *Strength*." (Ibid., pg. 3) The twenty-first, or final card is The World card. There can, also, be numerological keys of the Major Arcana, derived at from odd and even, and prime numbers.

With the Fool card, there are twenty-two cards in the Major Arcana.

> J. Ralston Skinner in *The Source of Measure* was convinced that the Pyramid was not a tomb, but a temple of initiation. He went further and linked the Pyramid to the Jewish cabala, a system of allegorical symbolism among the initiated which sets forth the secret teachings of the Bible, concealing the great cosmic principals of man's origin.
>
> According to Skinner the key to the cabala was said to be the geometrical relation of the area of the circle inscribed in the square, or the sphere in the cube. This gave rise to the relation of the diameter to the circumference of a circle, with the numerical value of the relation expressed in integrals, such as 22/7.
>
> The relation of diameter to circumference, says Skinner, was considered a supreme one, connected with the god names Elohim and Jehovah, the first being the circumference, the second the diameter, which were numerical expressions of these relations. (Tompkins, 1971, pp. 260-261)

22/7 is equal to 3.1428571 . . . , a very close approximation of π. Also, twenty-two is equal to two times eleven.

I

The Historical Development of the Golden Section
from Ancient Egypt

F. The Relationship between 1 2/3, the Golden Ratio, and π/2

1 2/3 is 1.$\bar{6}$. The golden ratio is 1.6180339 . . . π/2 is 1.5707963 . . .
The difference between 1 2/3 and the golden ratio is .04863277 . . . The
difference between the golden ratio and π/2 is .04723757 . . . The relative
difference between these three values is minuscule.

The continued fraction for 1 2/3 is

$$1\ 2/3 = 1 + \cfrac{1}{1 + \cfrac{1}{1 + \cfrac{1}{1}}}$$

which is a partial continued fraction of the golden ratio. 1 2/3 is, also, equal
to 5/3, which is the ratio of two consecutive Fibonacci numbers.

The continued fraction for π/2 is

$$\frac{\pi}{2} = 1 - \cfrac{1}{3 - \cfrac{2\cdot 3}{1 - \cfrac{1\cdot 2}{3 - \cfrac{4\cdot 5}{1 - \cfrac{3\cdot 4}{3 - \cfrac{6\cdot 7}{1 - \cfrac{5\cdot 6}{3 - \ddots}}}}}}}$$

$\pi/2$ may, also, be expressed as the infinite product,

$$\frac{\pi}{2} = \frac{2\cdot 2\cdot 4\cdot 4\cdot 6\cdot 6\cdot 8\cdot 8}{1\cdot 3\cdot 3\cdot 5\cdot 5\cdot 7\cdot 7\cdot 9} \; \cdots$$

    The ratio of the circumference of a semicircle of a circle to the diameter of the circle is $\pi/2$. This can be proven, as follows. See the diagram in Figure 8.

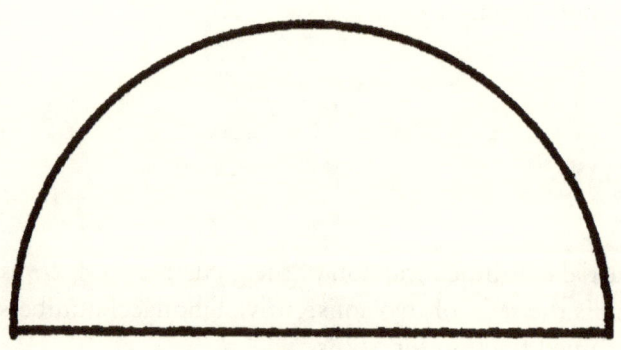

Figure 8

$$C_{ci} = 2\pi r$$
$$D_{ci} = 2r$$
$$\therefore \quad (1/2)C_{ci}/D_{ci} = \pi r/2r$$
$$= \pi/2$$

The figure of the semicircle of a circle and its diameter has aesthetic values of its own, which have been, and can be used in art, including drawing, painting, pottery, and sculpture, architecture, and materials' design. Figure 9 is a drawing of a fanlight, or semicircular window placed over a door, or window. According to *Webster's Third New International Dictionary of the English Language, Unabridged* (1993), a fanlight is "n: a semicircular window made with radiating sash bars like the ribs of a fan and placed over a door or window; *broadly*: a window over a door or window".

fanlight

Figure 9

Figure 10 is a photograph of the facade of San Miniato in Florence, Italy. Each of the three doors, or openings contain semicircular shapes above their tops.

Figure 10

Figure 11 is a photograph of the facade of the Saint Laurence Biblioteca, designed by Michelangelo, in Florence, Italy. Instead of semicircular shapes above every window, or door, Michelangelo has designed, alternately, isosceles triangles, which can be inscribed in semicircles.

Figure 11

Figure 12 is an artistic representation of day and night. In the circle are inscribed two, horizontal semicircles. It represents the wholeness arrived at by combing two opposites. In the symbol of the Tao, an ancient Chinese mandala, can, also, be found the design of the semicircle of a circle, and its diagonal.

Figure 12

The ratio of the area of a circle, inscribed inside a square, to the area of a 53° 7' 48", 63° 26' 58", 63° 26' 58" triangle, inscribed inside the same square, is $\pi/2$. This can be proven, as follows. See the diagram in Figure 13.

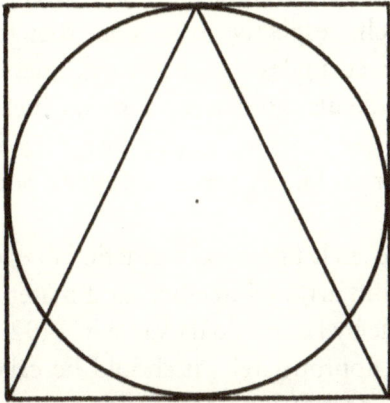

Figure 13

$$A_{ci} = \pi r^2$$
$$A_{tr} = (1/2) * 2r * 2r$$
$$= r * 2r$$
$$= 2r^2$$
$$\therefore \quad A_{ci}/A_{tr} = \pi r^2/2r^2$$

$$= \pi/2$$

The area of the inscribed, isosceles triangle is equal to one-half the area of the square. This can be proven, as follows. Consider the diagram in Figure 13. If the length of the sides of the square are one, then

$$A_{tr} = 1/2 * base * height$$
$$= (1/2) * 1 * 1$$
$$= 1/2$$
$$A_{sq} = 1 * 1$$
$$= 1$$
$$\therefore \quad A_{tr}/A_{sq} = (1/2)/1$$
$$= 1/2$$

If the length of the sides of the square are one, then the length of the diagonal of this square will be $\sqrt{2}$, ($\sqrt{1^2+1^2}$). The length of the two lateral sides of the inscribed isosceles triangle will be $\sqrt{5}/2$, ($\sqrt{(1/2)^2+1^2}$).

If an equivalent isosceles triangle is inscribed on each side of this square, and if each pair of lateral sides of these four triangles are extended horizontally, or vertically along each, respective base side, then two overlapping, and perpendicular root five rectangles will be formed. These are equivalent to four equivalent, and perpendicular golden rectangles, which overlap each other to the extent of a square. See Chapter VI., Section A., "The Root Rectangles". Also, Chapter III., Section D., "A Series of Golden Square Hypercubes, and the Fourth Dimension".

The squaring of the circle figure has aesthetic values of its own, which have been, and can be used in art, architecture, and materials' design. Figure 14 shows the drawing done by Leonardo da Vinci in 1492, called "The Cannons of Proportion". More appropriately, it should be called "The Cannons of Vitruvius", because the proportions, and design were taken from the work by Vitruvius, called *The Ten Books on Architecture*, published in 4 B.C. Vitruvius was an Italian architect.

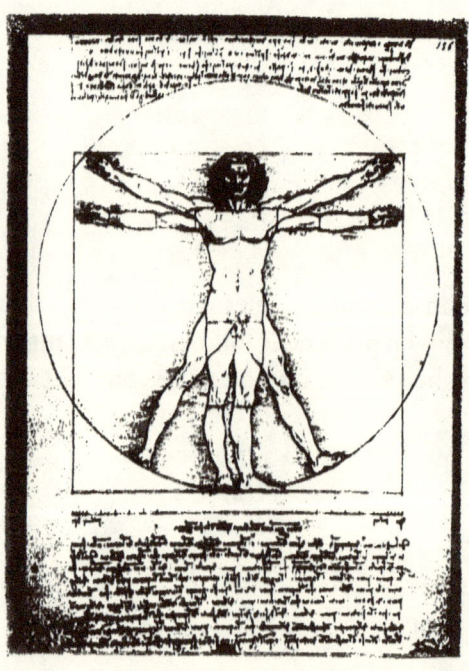

Figure 14

Figure 15 shows the squaring of the circle shape in a letter of the English alphabet.

Lucas Paciolus, a Minorite friar, wrote several books on mathematics. His De divina proportione, *contains material provided by an admirer, Leonardo da Vinci. When da Vinci's patron, Lodovico Sforza, commissioned the construction of a monumental bronze horse, da Vinci obtained the mathematics for the casting from Paciolus.*

Figure 15

In Brunes' reconstruction of the secret geometry, the cross emerges as the first geometric addition to the circle and square, and is the key not only to the solution of geometric problems but to the development of numerals and the alphabet.

By including the diagonals, every number both Latin and Arabic and all the letters of several alphabets may be obtained. (Tompkins, 1971, p. 261)

Figure 16 is a sixth century B.C. Greek, bronze mirror. The overall shape is a double square shape. AB and BC are squares. The rectangles BD and AF are .809 rectangles, or two golden rectangles (1.6180/2=.809). The rectangles CG and BH are both .191 rectangles, and are equivalent. The top square, and the circle of the mirror form the squaring of the circle shape.

Figure 16

Figure 17 is a seventeenth century drawing by Jamsthaler called, "All things do live in the three/ But in the four they merry be" (Jung, 1974, p. 199). It depicts the ultimate union of man and woman encircling the world. In the overall design plan, we can find the design theme of the squaring of the circle, and inscribed 53° 7' 48", 63° 26' 58", 63° 26' 58" triangle.

59. "All things do live in the three / But in the four they merry be." (Squa the circle.)—Jamsthaler, *Viatorium spagyricum* (1625)

Figure 17

The ratio of the area of a circle to the area of a square, inscribed in the circle is $\pi/2$. This can be proven, as follows. Consider the diagram in Figure 18.

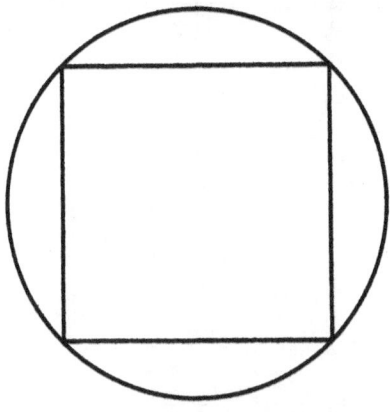

Figure 18

If the length of the sides of the square are one, then

$$A_{ci} = \pi r^2$$
$$= \pi(\sqrt{2}/2)^2$$
$$= \pi * 2/4, \text{ or } \pi/2$$
$$A_{sq} = 1 * 1$$
$$= 1$$
$$\therefore \; A_{ci}/A_{sq} = (\pi/2)/1$$
$$= \pi/2$$

The circling of the square figure has aesthetic values of its own, which have been, and can be used in art, architecture, and materials' design.

The diagrams in Figures 8, 13, and 18 can be combined to form a, perfectly, symmetrical pattern, as shown in Figure 19.

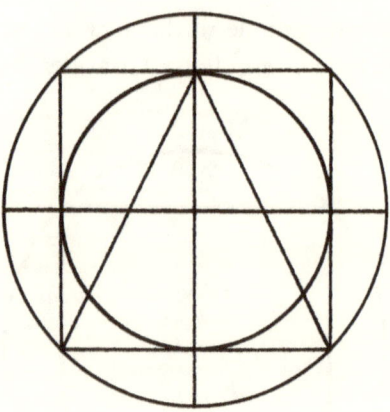

Figure 19

This figure is so beautiful, because all the parts are related to the whole through the same proportion, $\pi/2$.

One of the reasons that the Great Pyramid of Cheops is so beautiful is because the ratio of the length of its base side to its altitude is $\pi/2$.

The average of 1 2/3, and $\pi/2$ is (1.6666666+1.5707963)/2, or 1.6187315. This is a very close approximation to the golden ratio. For all practical purposes, 1 2/3 and $\pi/2$ can be considered as rough approximations of the golden ratio.

All of the Gothic mason's marks are based on either the semicircle and its diagonal, and/or the squaring of the circle, and/or the circling of the square figures. These mason marks were personal marks, and were used to be drawn and proved when the mason traveled to other mason's lodges.

CHAPTER II.

The Golden Triangle

II

The Golden Triangle

A. Introduction

The golden triangle is an isosceles triangle. It has been known, historically, as the Sublime Triangle, the Triangle of Plutarch, and the Triangle of the Pentalpha. The measure of its smallest angle is 36°. The measure of its two largest angles is 72°. Therefore, the measure of the smaller angle is equal to, exactly, one-half the measure of the two larger angles (see Figure 20). The golden triangle is a fundamental geometric structure, and is found in the geometries of the regular pentagon, and the regular decagon. It has been used, historically, in art and architecture throughout the ages, including during the classical ages of Egypt and Greece.

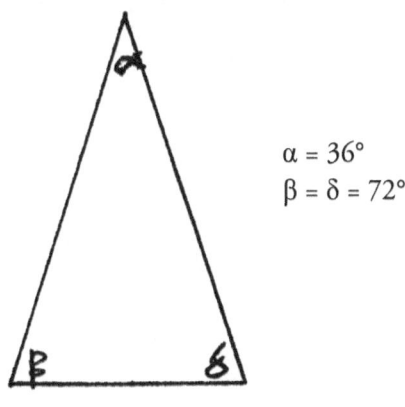

$\alpha = 36°$
$\beta = \delta = 72°$

Figure 20

The golden triangle is unique, because the ratio of the length of the two longer, or lateral sides to the length of the shorter, or base side is equal to the golden ratio.

The golden triangle has several applications to pure geometry. These include in the golden spiral of the golden triangle; two spirals made from golden triangles; three different types of towers, made from golden triangles; the orthocenter, the incenter, and the circumcenter; one golden triangle pair; a study of length, area, and volume; and the golden diamond.

It is found in such fundamental geometric structures as the regular pentagon, the pentagram, the regular decagon, the regular dodecahedron, the regular icosahedron, the stellated dodecahedron, and the stellated icosahedron. Because many of these structures were either known by, or discovered by the Pythagoreans, it is known that the golden triangle was studied by the Pythagoreans.

A golden triangle can be formed from, and can be inscribed inside a regular pentagon. Any triangle formed by two adjacent diagonals, and one side of a regular pentagon is a golden triangle. This can be proven, as follows. Consider the diagram in Figure 21.

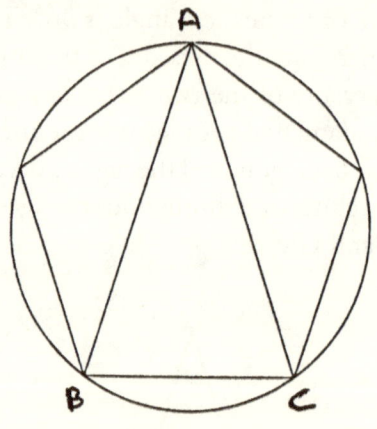

Figure 21

$$\angle BAC = (1/2)(1/5)360°$$
$$= (1/2)72°$$
$$= 36°$$
$$\angle ABC = \angle ACB = (1/2)(2/5)360°$$
$$= (1/2)144°$$
$$= 72°$$

Triangle ABC is a 36°, 72°, 72° triangle. Therefore, it is a golden triangle. Five different golden triangles can be inscribed inside of a regular pentagon. If they are, a pentagram is formed.

The pentagram was used as a symbol of secrecy in the Pythagorean brotherhood. The pentagram is, also, found in the Tarot cards of modern day. For example, one of the four suites of the Minor Arcana is pentacles, which stands for money, or the merchant class, during the medieval ages. It is a symbol of a circle with a pentagram inscribed in it. Also, as the symbol of one of these suites, it is found on the Magician card, which is the first card of the Major Arcana. In the realm of the occult, or mysticism, the symbol of an upside down pentagram represents the devil, or evil. For example, on the number fifteen card of the Major Arcana, the Devil, the symbol of an inverted pentagram is found placed between the horns of the Devil.

A golden triangle can be formed from, and forms one of the symmetric partitions of a regular decagon. Any triangle formed by two consecutive radii, and one side of a regular decagon is a golden triangle. This can be proven, as follows. Consider the diagram in Figure 22.

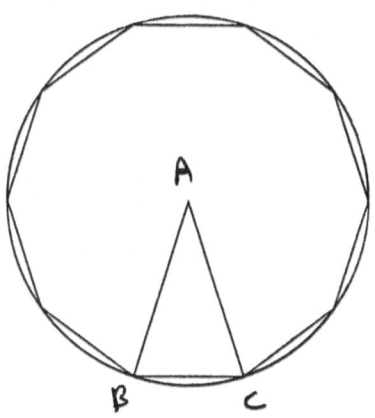

Figure 22

$$\angle BAC = (1/10)360°$$
$$= 36°$$
$$\angle ABC = \angle ACB = (1/2)(4/10)360°$$
$$= (1/2)144°$$
$$= 72°$$

Triangle ABC is a 36°, 72°, 72° triangle. Therefore, it is a golden triangle. In any regular decagon, the ten successive radii form ten successive golden triangles, which, symmetrically, completely, fill the boundaries of the regular decagon.

The purpose of the present study is to elaborate on various geometric constructions, which depend on, or involve the concept of the golden triangle. The ramifications of the golden triangle found in art and architecture will be explored in chapter six.

II

The Golden Triangle

B. The Golden Spiral of the Golden Triangle

The golden spiral of the golden triangle is a spiral, which is formed by the constant subdivision of golden triangles into a smaller golden triangle, and a 36°, 36°, 108° triangle in either a clockwise, or counterclockwise fashion in a continuous manner within the golden triangle.

In any golden triangle, another smaller golden triangle, and a 36°, 36°, 108° triangle can be constructed by drawing in one angle bisector at either base vertex of the golden triangle. The angle formed on either side of the angle bisector at the base vertex, that it transects, will be 36°. The opposite base vertex of the original golden triangle will be 72°. The remaining internal angle of the triangle, formed by the base of the original golden triangle, the angle bisector, and a segment of the respective lateral side, which is transected by the angle bisector, is 72°. This can be proven, as follows. Consider the diagram in Figure 23.

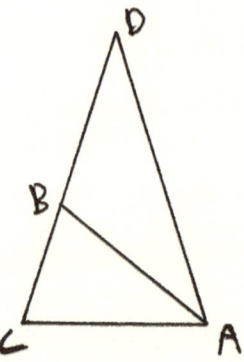

Figure 23

$$180° = \angle ABC + \angle BCA + \angle CAB$$
$$\angle ABC = 180° - \angle BCA - \angle CAB,$$
$$\text{but } \angle BCA = 72°, \text{ and}$$
$$\angle CAB = 36°$$
$$\therefore \quad \angle ABC = 180° - 72° - 36°$$
$$= 180° - 108°$$
$$= 72°$$

Therefore, triangle ABC is a 36°, 72°, 72° triangle, or another golden triangle. Therefore, $\overline{AC}$ is equal in length to $\overline{AB}$, and the ratio of the lengths of $\overline{AB}$ to $\overline{BC}$ is equal to the golden ratio.

Also, triangle ABD is a 36°, 36°, 108° triangle. This can be proven, as follows. Consider the diagram in Figure 23.

$$\angle ABD = 180° - \angle ABC$$
$$= 180° - 72°$$
$$= 108°$$
$$\angle BAD = 36°, \text{ and}$$
$$\angle BDA = 36°$$

Therefore, triangle ABD is a 36°, 36°, 108° triangle.

In any 36°, 36°, 108° triangle, the ratio of the length of the longer side, or base to the length of the two shorter sides is equal to the golden ratio. This can be proven, as follows. Consider the diagram in Figure 24.

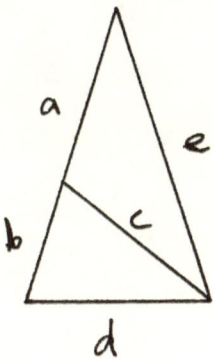

Figure 24

$$e = a+b$$
$$d = c = a$$
$$c/b = \tau$$
$$\therefore \quad a/b = \tau$$
$$\therefore \quad (a+b)/a = \tau$$
$$\therefore \quad e/a = e/c = \tau$$

The golden spiral of the golden triangle is a spiral, which is formed by joining together with line segments the successive orthocenters, incenters, or circumcenters of the successive golden triangles.

In any golden triangle, the spiral pole can be found by determining the point of intersection of two corresponding medians within any two consecutive golden triangles. See Figure 25. The medians of a triangle are the line segments, drawn from the three vertices of the triangle to the respective, opposite sides, and bisecting those opposite sides. The spiral pole can be formed on either the right-hand, or the left-hand side of the golden triangle. The golden spiral of the golden triangle consists of an infinite number of golden triangles, which become, infinitely, small, as they approach the spiral pole.

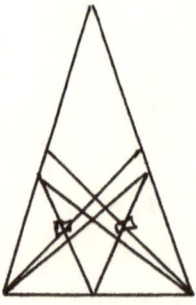

Figure 25

Because the ratio of the lengths of the larger to the smaller of similar parts of consecutive golden triangles is equal to the golden ratio, the ratio of the lengths of the larger to the smaller of consecutive chords of this spiral is equal to the golden ratio. Also, because the arcs of the spiral between each pair of consecutive radii is similar, the ratio of the lengths of the larger to the smaller of consecutive arcs of this spiral is equal to the golden ratio. Therefore, this spiral is called the golden spiral of the golden triangle. See Figure 26.

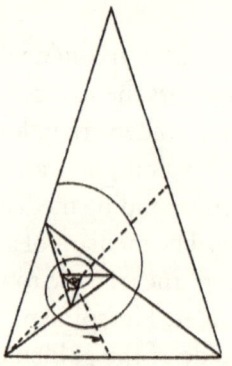

Figure 26

Because this spiral grows, constantly, in size, but without changing in shape, it is a logarithmic spiral.

II

The Golden Triangle

C. Two Types of Spirals, Made from Golden Triangles

Two different types of spirals, made from golden triangles, can be constructed. The first can be constructed by arranging a series of golden triangles around a spiral pole, so that one of the 72° base angles of each golden triangle touches the spiral pole. Then, one of the lateral sides of each golden triangle forms the base, or shorter side of each larger, consecutive golden triangle. See Figure 27.

The first type of spiral, made from golden triangles, is equivalent to a spiral, made from a series of radii, rotated around a spiral pole at an angle of 72°, and whose successive lengths are in golden ratio. Then, the ratio of the lengths of the larger to the smaller of consecutive radii will be equal to the golden ratio. Also, the ratio of the lengths of the larger to the smaller of consecutive chords will be equal to the golden ratio.

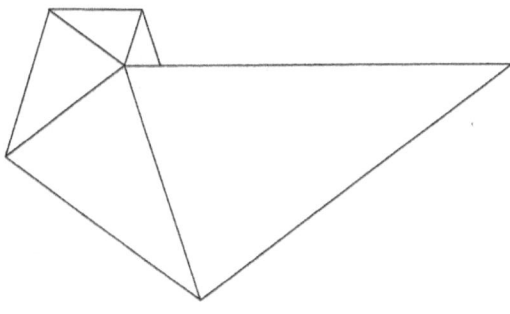

Figure 27

The second type of spiral of golden triangles can be constructed from a series of 36°, 36°, 108° triangles, which are rotated by one 36° angle around a spiral pole. Then, the base, or longer side of each 36°, 36°, 108° triangle will form one of the shorter sides of each larger, consecutive 36°, 36°, 108° triangle. See Figure 28.

The second type of spiral, made from golden triangles, is equivalent to a series of line segments, rotated at a 36° angle around a spiral pole, and whose successive lengths are in golden ratio. Then, the ratio of the lengths of the larger to the smaller of consecutive radii will be equal to the golden ratio. Also, the ratio of the lengths of the larger to the smaller of consecutive chords will be equal to the golden ratio.

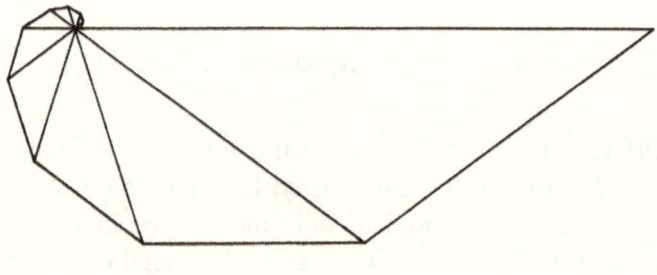

Figure 28

Both the first, and the second types of spirals, made from golden triangles, are equiangular, or logarithmic spirals, because each golden triangle, or 36°, 36°, 108° triangle grows in size, but remains constant in shape.

II

The Golden Triangle

D. Three Different Types of Towers, Made from Golden Triangles

Three different types of towers, made from golden triangles, exist. The first can be constructed by arranging a series of line segments inside a golden triangle, so that they alternate back and forth from one side of the golden triangle to the other, and so that they all lie at an angle of 36° from the horizontal. See Figure 29. In this tower, each triangle formed by two consecutive line segments inside the golden triangle, and a segment of either of the two lateral sides of the external golden triangle is another golden triangle. This can be proven, as follows. Consider the diagram in Figure 29.

The length of the first line segment is equal to the length of the base of the original golden triangle. Also, the base vertex angle between the first line segment, and the base of the original golden triangle is 36°. Therefore, the triangle formed by the first line segment, the base, and a segment of one of the lateral sides is another golden triangle. In each triangle, thereafter, the measure of the smallest angle is 36°. The measure of the two largest angles is 72°. Therefore, these triangles are, also, golden triangles.

Therefore, this type of tower of golden triangles consists of an infinite number of golden triangles, which become, infinitely, small at the apex of the original golden triangle.

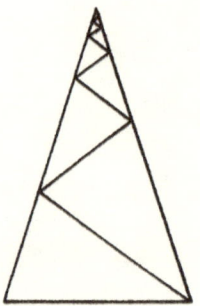

Figure 29

Because the first line segment at the base of the golden triangle can be created at either the right-hand, or left-hand side, this type of tower, made from golden triangles, can be either right or left-handed.

A golden triangle screw can be constructed from the first type of tower, made from golden triangles. This screw can be constructed by superimposing the first type of tower, made from golden triangles, on a right circular cone, whose vertical cross section yields a golden triangle. See Figure 30.

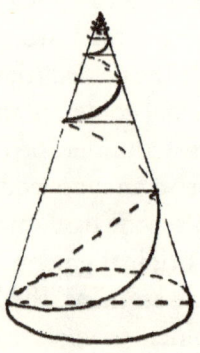

Figure 30

Each line segment of the first type of tower, made from golden triangles, lies at an angle of 36° from the base. Therefore, the tangent to the curve of the screw at every point on the curve, also, lies at an angle of 36° from the base.

The second type of tower, made from golden triangles, can be constructed from the first type of tower, made from golden triangles. By superimposing a right-handed first type of tower, made from golden triangles, on top of a left-handed first type of tower, made from golden triangles, then this type of tower, made from golden triangles, will be constructed. See Figure 31.

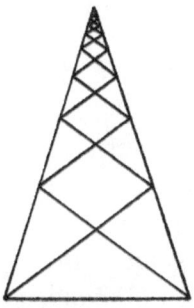

Figure 31

This type of tower, made from golden triangles, can, also, be called a golden triangle Christmas tree.

The third type of tower, made from golden triangles, can be constructed from the second type of tower, made from golden triangles. By drawing in horizontal, and parallel struts between the symmetric endpoints of the line segments of the second type of tower, made from golden triangles, on the two lateral sides of the original golden triangle, the third type of tower made from golden triangles can be constructed. See Figures 32, a and 32, b.

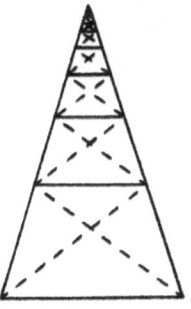

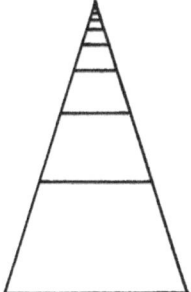

Figure 32, a   Figure 32, b

This type of tower, made from golden triangles, has several properties. One of these is that each triangle, formed by one horizontal strut, and two equal segments of the lateral sides of the original golden triangle is another golden triangle. This is so, because each horizontal strut is parallel to the base of the original golden triangle, and its lateral sides are segments of the lateral sides of the original golden triangle.

Also, the ratio of the lengths of the respective sides of the larger to the smaller of two consecutive golden triangles in this tower is equal to the golden ratio. This can be proven, as follows. Consider the diagram in Figure 32, b. The lengths of the bases of consecutive triangles will be in golden ratio. Therefore, the lengths of the respective sides of consecutive golden triangles will, also, be in golden ratio. Also, the ratio of the lengths of the larger to the smaller of two consecutive segments on either lateral side of the original golden triangle is equal to the golden ratio.

Therefore, this tower made from golden triangles consists of an infinite number of golden triangles, which become, infinitely, small, as the series approaches the apex.

II

The Golden Triangle

E. The Orthocenter, the Incenter, and the Circumcenter

The orthocenter of a triangle is the point in the center of the triangle, where the three altitudes meet (see Figure 33). It can be either inside, or outside the triangle, depending on the type of triangle.

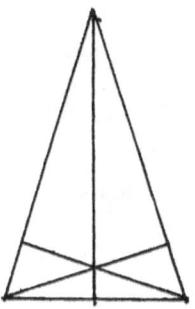

Figure 33

The incenter of a triangle is the point in the center of the triangle, where the three angle bisectors meet. It is called the incenter, because it is, also, the center of the circle, which is inscribed inside the triangle, which is called the incircle (see Figure 34). It must always be inside the triangle. Also, because all radii of a circle are perpendicular to the line segments, which form their respective tangents, the three radii of the incircle, which, also, touch the three sides of the triangle, are perpendicular to these respective sides (see Figure 35).

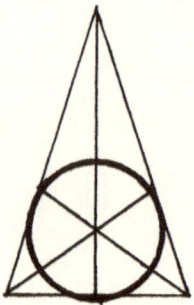

Figure 34

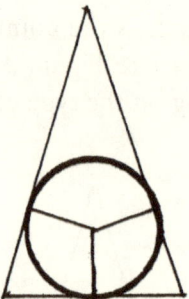

Figure 35

The circumcenter of a triangle is the point in the center of the triangle, which is the center of the circumcircle, which circumscribes the triangle (see Figure 36). It can be either inside, or outside the triangle, depending on the type of triangle. The three line segments, which go from the respective vertices of the triangle to the circumcenter are called circumradii. All the circumradii of a triangle are of an equal length.

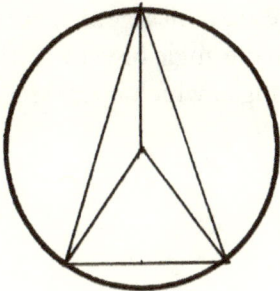

Figure 36

In this section, I prove that the angle between the base of a golden triangle, and one of its base altitudes is 18°; the angle between the base of a golden triangle, and one of its base angle bisectors is 36°; and the angle between the base of a golden triangle, and one of its base circumradii is 54°. Thus, proving that the angle between each of these line segments is 18°. Thus, proving a fundamental geometric correlation between these three points.

The angle between the base of a golden triangle, and one of the base altitudes is 18°. This can be proven, as follows. Consider the diagram in Figure 37.

$$\angle\beta = 90°$$
$$\angle\delta = 72°$$
$$180° = \angle\alpha + \angle\beta + \angle\delta$$
$$\text{Therefore, } \angle\alpha = 180° - \angle\beta - \angle\delta$$
$$= 180° - 90° - 72°$$
$$= 90° - 72°$$
$$= 18°$$

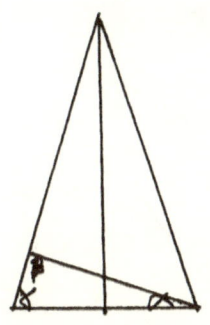

Figure 37

The angle between the base of a golden triangle, and one of the base angle bisectors is 36°. Since the angle bisector at the base of a golden triangle bisects the respective base angle, which it transects, the measure of this angle is 36° (see Figure 38).

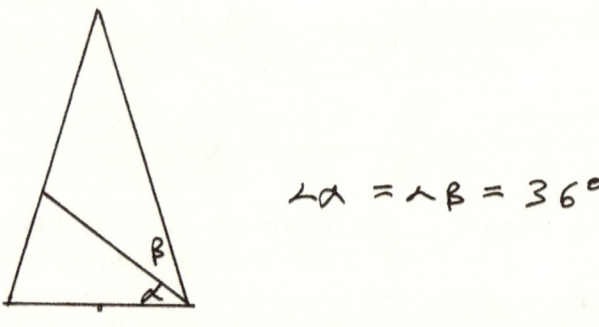

$\angle \alpha = \angle \beta = 36°$

Figure 38

The angle between the base of a golden triangle, and one of the base circumradii is 54°. This can be proven, as follows. Consider the diagram in Figure 39. The angle, formed by the two base circumradii, cuts out a chord, which subtends an arc of one-fifth of the circumcircle. Therefore, the measure of this angle is equal to the number of degrees in one-fifth of a circle, or 72°. Since the triangle formed by the two base circumradii, and the base of the original golden triangle is an isosceles triangle, the measure of the two base angles are equal. They are 54°.

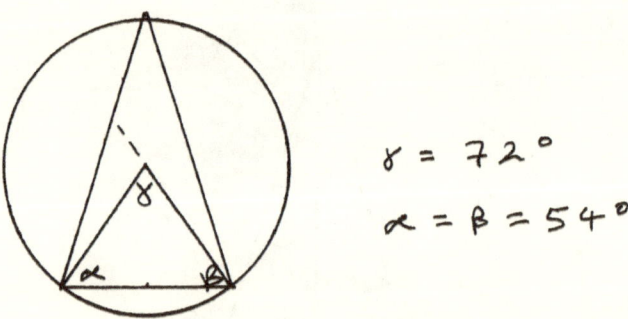

$\gamma = 72°$

$\alpha = \beta = 54°$

Figure 39

Also, the line segment, which goes from one base vertex of a golden triangle to the respective, opposite endpoint of the second strut from the bottom of the third type of tower made from golden triangles goes through the circumcenter. This can be proven, as follows. Consider the diagram in Figure 40.

$$\angle DAB = \angle GAD = 36°$$
$$\overline{AD} = \overline{AB}$$
$$\overline{AG}/\overline{AE} = \tau$$
$$\overline{AG}/\overline{AB} = \tau$$
$$\therefore \ \ \overline{AE} = \overline{AB}$$
$$\therefore \ \ \overline{AE} = \overline{AD}$$
$$\overline{FG}/\overline{DF} = \tau$$
$$\overline{FG}/\overline{EF} = \tau$$
$$\therefore \ \ \overline{DF} = \overline{EF}$$

Therefore, quadrilateral AEFD consists of two pairs of adjacent, equivalent sides. And, the diagonal $\overline{AF}$ bisects the two opposite angles, that it transects, of the quadrilateral AEFD, $\angle EFD$ and $\angle DAE$. Therefore, the measure of angle FAD is 18°. Therefore, the measure of angle FAB is 54°. This is the same angle, as the angle at which one circumradii at the base of a golden triangle lies from the base of the golden triangle.

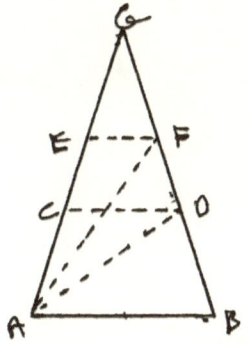

Figure 40

The relationship between the altitudes, the angle bisectors, the circumradii, the first two struts from the bottom of the third type of tower, made from golden triangles, the incircle, and the circumcircle of a golden triangle is shown in Figure 41.

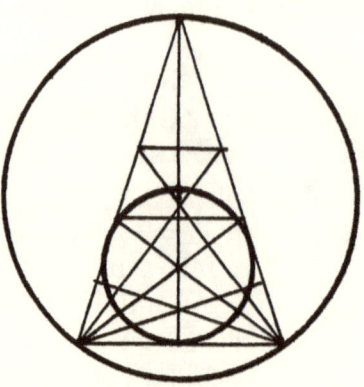

Figure 41

II

The Golden Triangle

F. One Golden Triangle Pair

A golden triangle pair consists of concentric golden triangles with the respective sides being parallel. The ratio of the lengths of the parallel sides is equal to either a multiple, a power, or a combination of multiples and powers of the golden ratio.

A golden triangle pair with a common incenter, and with the ratio of the lengths of the parallel sides being equal to τ^2 is depicted in Figure 42.

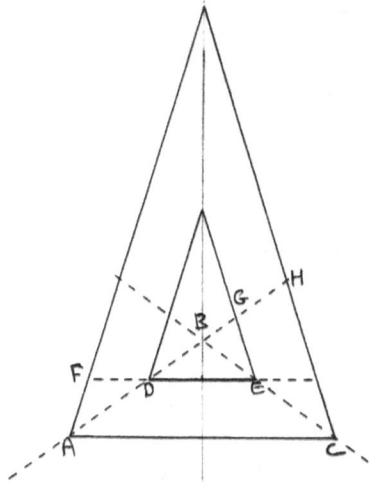

Figure 42

This golden triangle pair has some interesting properties. One property is that the length of the two angle bisectors, drawn from the base vertices of the external golden triangle to the, respective, opposite sides, is in golden ratio to the length of the two line segments, drawn from the base vertices of the external golden triangle to the incenter. This can be proven, as follows. Consider the diagram in Figure 42. Triangle ABC is a 36°, 36°, 108° triangle with the length of the longer side being in golden ratio to the length of the two shorter sides. Therefore,

$$\overline{AC}/\overline{AB} = \tau, \text{ but}$$
$$\overline{AC} = \overline{AH}$$
$$\therefore \quad \overline{AH}/\overline{AB} = \tau$$

A second property is that the length of the two line segments, drawn from the base vertices of the external golden triangle to the incenter, is in golden ratio to the length of the two line segments, drawn from the incenter to the, respective, opposite sides of the external golden triangle. This can be proven, as follows. Consider the diagram in Figure 42.

$$\overline{AH}/\overline{AB} = \tau$$
$$\therefore \quad \overline{AB}/\overline{BH} = \tau$$

These same two properties hold true for the internal golden triangle.

$$\overline{DG}/\overline{DB} = \tau, \text{ and}$$
$$\overline{DB}/\overline{BG} = \tau$$

A third property is that the length of the two line segments, drawn from the base vertices of the external golden triangle to the, respective, base vertices of the internal golden triangle, is in golden ratio to the length of the two line segments, drawn from the base vertices of the internal golden triangle to the incenter. This can be proven, as follows. Consider the diagram in Figure 42. Because the ratio of the lengths of the parallel sides is equal to τ^2, the ratio of the lengths of similar line segments within the two golden triangles is, also, equal to τ^2. That is,

$$\overline{AB}/\overline{DB} = \tau^2, \text{ but}$$
$$\overline{AB} = \overline{AD} + \overline{DB}$$

$$\therefore \ (\overline{AD} + \overline{DB})/\overline{DB} = \tau^2$$
$$\overline{AD}/\overline{DB} + \overline{DB}/\overline{DB} = \tau^2$$
$$\overline{AD}/\overline{DB} + 1 = \tau^2$$
$$\therefore \ \overline{AD}/\overline{DB} = \tau^2 - 1$$
$$= \tau$$

A fourth property is that the length of the two line segments, drawn from the base vertices of the external golden triangle to the incenter, is in golden ratio to the length of the two line segments, drawn from the base vertices of the external golden triangle to the, respective, base vertices of the internal golden triangle. This can be proven, as follows. Consider the diagram in Figure 42.

$$\overline{AD}/\overline{DB} = \tau$$
$$\therefore \ \overline{AB}/\overline{AD} = \tau$$

A fifth property is that the length of the two line segments, drawn from the incenter to the, respective, opposite sides of the external golden triangle, is in golden ratio to the length of the two line segments, drawn from the base vertices of the internal golden triangle to the incenter. This can be proven, as follows. Consider the diagram in Figure 42.

$$\overline{AB}/\overline{BH} = \tau$$
$$\overline{AB}/\overline{AD} = \tau$$
$$\therefore \ \overline{BH} = \overline{AD}$$
$$\overline{AD}/\overline{DB} = \tau$$
$$\therefore \ \overline{BH}/\overline{DB} = \tau$$

A sixth property is that the length of the two line segments, drawn from the lateral sides of the internal golden triangle to the, respective, lateral sides of the external golden triangle, is in golden ratio to the length of the two line segments, drawn from the incenter to the, respective, opposite sides of the internal golden triangle. This can be proven, as follows. Consider the diagram in Figure 42.

$$\overline{BH}/\overline{DB} = \tau$$
$$\overline{DB}/\overline{BG} = \tau$$
$$(\overline{BH}/\overline{DB})(\overline{DB}/\overline{BG}) = \tau^2$$

$$\therefore \quad \overline{BH}/\overline{BG} = \tau^2, \text{ but}$$
$$\overline{BH} = \overline{BG} + \overline{GH}$$
$$\therefore \quad (\overline{BG}+\overline{GH})/\overline{BG} = \tau^2$$
$$\overline{BG}/\overline{BG} + \overline{GH}/\overline{BG} = \tau^2$$
$$1 + \overline{GH}/\overline{BG} = \tau^2$$
$$\therefore \quad \overline{GH}/\overline{BG} = \tau^2 - 1$$
$$= \tau$$

A seventh property is that the length of the two line segments, drawn from the incenter to the, respective, opposite sides of the external golden triangle, is in golden ratio to the length of the two line segments, drawn from the lateral sides of the internal golden triangle to the respective lateral sides of the external golden triangle. This can be proven, as follows. Consider the diagram in Figure 42.

$$\overline{GH}/\overline{BG} = \tau$$
$$\therefore \quad \overline{BH}/\overline{GH} = \tau$$

An eighth property is that the length of the two angle bisectors, drawn from the base vertices of the internal golden triangle to the, respective, opposite sides of the internal golden triangle, is in golden ratio to the length of the two line segments, drawn from the lateral sides of the internal golden triangle to the, respective, lateral sides of the external golden triangle. This can be proven, as follows. Consider the diagram in Figure 42.

$$\overline{GH}/\overline{BG} = \tau$$
$$\overline{DB}/\overline{BG} = \tau$$
$$\therefore \quad \overline{GH} = \overline{DB}$$
$$\therefore \quad \overline{DG}/\overline{GH} = \tau$$

A ninth property is that the length of the base of the internal golden triangle is in golden ratio to the length of the two line segments, drawn from the base vertices of the internal golden triangle to the, respective, lateral sides of the external golden triangle. This can be proven, as follows. Consider the diagram in Figure 42. Triangles AFD and DBE are both 36°, 36°, 108° triangles with the length of the longer side being in golden ratio to the length of the two shorter sides. Therefore,

$$\overline{DE}/\overline{DB} = \tau, \text{ and}$$
$$\overline{AD}/\overline{FD} = \tau, \text{ but}$$
$$\overline{AD}/\overline{DB} = \tau$$
$$\therefore \quad \overline{DE} = \overline{AD}, \text{ and}$$
$$\overline{DE}/\overline{FD} = \tau$$

Other golden triangle pairs have, similarly, interesting properties, but they shall not be discussed here.

II

The Golden Triangle

G. A Study of Length, Area, and Volume

A series of golden triangles with a common circumcenter, and with the lengths of the parallel sides of consecutive golden triangles, being in golden ratio, is shown in Figure 43.

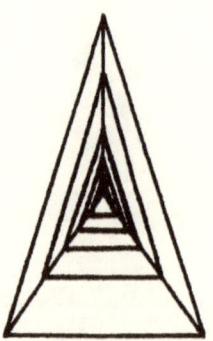

Figure 43

This series of golden triangles consists of an infinite number of golden triangles, which become, infinitely, dense at the center of logarithmic growth.

All the vertices of these golden triangles lie along the three circumradii, which extend from the three vertices of the largest golden triangle to the circumcenter. These vertices divide each of these circumradii into a series of line segments with the lengths of consecutive line segments, being in golden ratio.

A spiral of line segments, based on a series of golden triangles with a common circumcenter, and with the lengths of the parallel sides of consecutive golden triangles, being in golden ratio, is shown in Figure 44.

Figure 44

Two other series of golden triangles, one with a common incenter, and the other with a common orthocenter, with the lengths of the parallel sides of consecutive golden triangles, being in golden ratio, also, exist, but we will not consider them here. These series have the same properties, as the series in Figure 43.

A series of golden triangles with a common circumcenter, and with the areas of consecutive golden triangles, being in golden ratio is shown in Figure 45.

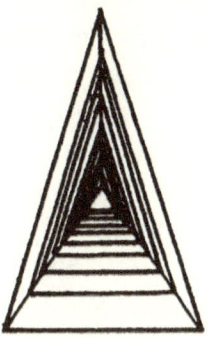

Figure 45

This series of golden triangles consists of an infinite number of golden triangles, which become, infinitely, dense at the center of logarithmic growth.

All the vertices of these golden triangles lie along the three circumradii, that extend from the three vertices of the largest golden triangle to the circumcenter. These vertices divide each of these circumradii into a series of line segments with the ratio of the lengths of consecutive line segments, being equal to $\tau^{1/2}$.

This series has two important properties. One property is that the ratio of the area of each golden triangle to the area between the golden triangle, and each consecutive, and larger golden triangle is equal to the golden ratio. This can be proven, as follows. Consider the diagram in Figure 46.

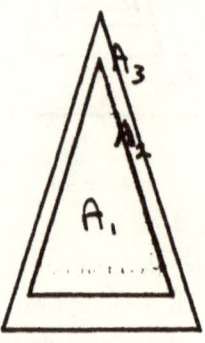

Figure 46

If A_1 = area of the smallest golden triangle,
A_2 = area between the smallest, and the largest golden triangle, and
A_3 = area of the largest golden triangle, then

$$A_3/A_1 = \tau$$
$$A_3 = A_1 + A_2$$
$$\therefore \ (A_1 + A_2)/A_1 = \tau$$
$$A_1/A_1 + A_2/A_1 = \tau$$
$$1 + A_2/A_1 = \tau$$
$$\therefore \ A_2/A_1 = \tau - 1$$
$$= 1/\tau$$
$$\therefore \ A_1/A_2 = \tau$$

This property holds true for each pair of consecutive golden triangles in this series.

The second property is that the ratios of contiguous areas between successive golden triangles is equal to the golden ratio. This can be proven, as follows. Consider the diagram in Figure 47.

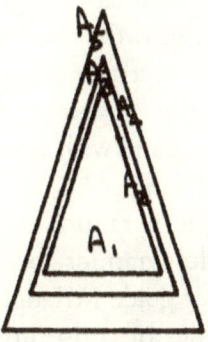

Figure 47

If A_1 = area of the smallest golden triangle

A_2 = area between the smallest, and the middle golden triangle

A_3 = area of the middle golden triangle

A_4 = area between the middle, and the largest golden triangle, and

A_5 = area of the largest golden triangle, then

$$A_1/A_2 = \tau$$
$$A_3/A_4 = \tau$$
$$A_3 = A_1 + A_2$$
$$\therefore (A_1 + A_2)/A_4 = \tau$$
$$A_1 + A_2 = \tau * A_4$$
$$(A_1 + A_2)/A_2 = \tau * (A_4/A_2)$$
$$A_1/A_2 + A_2/A_2 =$$
$$\therefore \tau + A_2/A_2 =$$
$$\tau + 1 =$$
$$\therefore \tau/\tau + 1/\tau = A_4/A_2$$
$$1 + 1/\tau =$$
$$\therefore 1 + (\tau - 1) =$$
$$\tau = \quad , \text{ or}$$
$$A_4/A_2 = \tau$$

This property holds true for every pair of contiguous areas between successive golden triangles in this series.

Two other series of golden triangles, one with a common incenter, and the other with a common orthocenter with the areas of consecutive golden triangles being in golden ratio, also, exist, but we will not consider them here. These series have the same properties, as the series in Figure 45.

These properties hold true for a series of golden triangle pyramids with a common center of logarithmic growth, and with the ratios of the volumes of consecutive golden triangle pyramids being equal to the golden ratio, except with volume substituted for area. A golden triangle pyramid is a pyramid with a regular polygon, as a base, and with a corresponding number of lateral faces, all of which are golden triangles.

A series of golden triangle pyramids with the ratios of the volumes of consecutive golden triangle pyramids, being equal to the golden ratio, completely, overlaps a series of golden triangle pyramids with the ratios of the lengths of the parallel sides of consecutive golden triangle pyramids, being equal to the golden ratio, if the two series have a common center of logarithmic growth, and if the two series are superimposed. The reason for this is, as follows. The ratio of the lengths of the parallel sides of successive golden triangle pyramids in the former series is equal to $\tau: \tau^{2/3}: \tau^{1/3}: 1$. The ratio of the lengths of the parallel sides of successive golden triangle pyramids in the latter series is equal to $\tau:1$. A side of respective length τ will, always, overlap another side of respective length τ. A side of respective length 1 will, always, overlap another side of respective length 1.

A series of golden triangle pyramids with the ratios of the volumes of consecutive golden triangle pyramids, being equal to the golden ratio, only overlaps every other golden triangle pyramid in a series of golden triangle pyramids with the ratios of the areas of the parallel faces of consecutive golden triangle pyramids, being equal to the golden ratio, if the two series have a common center of logarithmic growth, and if the two series are superimposed. The reason for this is, as follows. The ratio of the lengths of the parallel sides of successive golden triangle pyramids in the former series is equal to $\tau: \tau^{2/3}: \tau^{1/3}: 1$. The ratio of the lengths of the parallel sides of successive golden triangle pyramids in the latter series is equal to $\tau: \tau^{1/2}: 1$. A side of respective length τ will, always, overlap another side of respective length τ. A side of respective length 1 will, always, overlap another side of respective length 1. However, a side of respective length $\tau^{1/2}$ will never overlap a side of respective length τ, $\tau^{2/3}$, $\tau^{1/3}$, or 1.

CHAPTER III.

The Golden Square

III

The Golden Square

A. Introduction

A golden square consists of a pair of concentric squares with parallel sides, and with the sides of the internal square extending to the sides of the external square. The ratio of the lengths of the larger to the smaller of the parallel sides can be equal to either a multiple, a power, or a combinations of multiples and powers of the golden ratio.

The geometry of the golden square is fundamental to the geometry of the golden section. It plays an important part in the geometric breakdown of the golden rectangle into further golden rectangles, and golden squares. Also, it has aesthetic values of its own, which have been, and can be used in art (including drawing, painting, pottery, and sculpture), architecture, and materials' design. For instance, in an analysis of the proportions of the bones of the human body, the golden square can be found in at least one case.

The purpose of the present study is to elaborate on various geometrical constructions, which depend on, or involve the concept of a golden square.

III

The Golden Square

B. One Golden Square

In the book, *The Fibonacci Numbers* (1963, p. 44), by N. N. Vorobyov, the following problem is stated: "Let us note that if the golden rectangle I and the squares II and III are inscribed in a square, as shown in Fig. 8 (Figure 48), the remaining rectangle will likewise be a golden rectangle. The proof of this statement is left to the reader." The solution to this problem can be found, as follows. Consider the diagram in Figure 48.

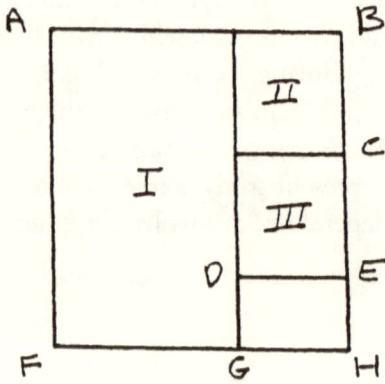

Figure 48

$$\overline{AF}/\overline{FG} = \tau$$
$$\overline{FH} = \overline{AF}$$
$$\therefore \quad \overline{FH}/\overline{FG} = \tau, \text{ and}$$
$$\overline{FG}/\overline{GH} = \tau, \text{ but}$$
$$\overline{BH} = \overline{FH}$$
$$\overline{BC} = \overline{GH}, \text{ and}$$
$$\overline{BH}-\overline{BC} = \overline{FH}-\overline{GH}$$
$$= \overline{FG} = \overline{CH}$$
$$\therefore \quad \overline{CH}/\overline{BC} = \tau, \text{ but}$$
$$\overline{CE} = \overline{BC}$$
$$\therefore \quad \overline{CH}/\overline{CE} = \tau, \text{ and}$$
$$\overline{CE}/\overline{EH} = \tau, \text{ but}$$
$$\overline{DE} = \overline{CE}$$
$$\therefore \quad \overline{DE}/\overline{EH} = \tau$$

Therefore, rectangle DEHG is another golden rectangle.

If the components of the square shown in Figure 48 are rearranged, so that golden rectangle DEHG is placed between squares II and III, instead of below them, then the square shown in Figure 49 will result.

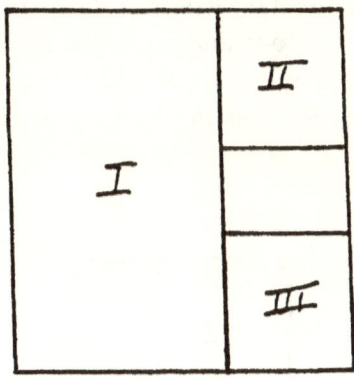

Fig. 49

By constructing symmetrical line segments on each side of the square shown in Figure 49, a golden square will result, as shown in Figure 50.

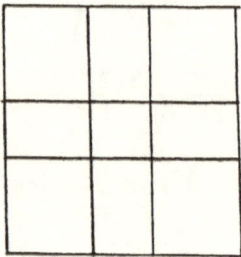

Figure 50

This golden square consists of a pair of concentric squares with parallel sides, and with the sides of the internal square extending to the sides of the external square. The ratio of the lengths of the parallel sides is equal to τ^3.

This golden square consists of nine symmetrical components: four squares, one in each corner; four golden rectangles, one between each pair of corner squares; and a square in the center.

The diagonal of this golden square has some interesting properties. One property is that the the ratio of the length of the two diagonals of the external square to the length of the four line segments, drawn from the vertices of the internal square to the respective opposite vertices of the external square, is equal to the golden ratio. This can be proven, as follows. Consider the diagram in Figure 51.

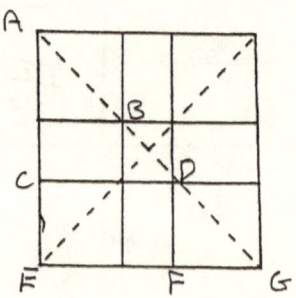

Figure 51

$$\Delta AEG \backsim \Delta ACD$$
$$\overline{EG}/\overline{EF} = \tau, \text{ but}$$
$$\overline{EF} = \overline{CD}$$
$$\therefore \quad \overline{EG}/\overline{CD} = \tau$$
$$\therefore \quad \overline{AG}/\overline{AD} = \tau$$

A second property is that the ratio of the length of the four line segments, drawn from the vertices of the internal square to the respective opposite vertices of the external square, to the length of the four line segments, drawn from the vertices of the internal square to the respective vertices of the external square, is equal to the golden ratio. This can be proven, as follows. Consider the diagram in Figure 51.

$$\overline{AG}/\overline{AD} = \tau$$
$$\therefore \quad \overline{AD}/\overline{DG} = \tau, \text{ but}$$
$$\overline{AB} = \overline{DG}$$
$$\therefore \quad \overline{AD}/\overline{AB} = \tau$$

A third property is that the ratio of the length of the four line segments, drawn from the vertices of the internal square to the respective vertices of the external square, to the length of the two diagonals of the internal square is equal to the golden ratio. This can be proven, as follows. Consider the diagram in Figure 51.

$$\overline{AD}/\overline{AB} = \tau$$
$$\therefore \quad \overline{AB}/\overline{BD} = \tau$$

Other golden squares have, similarly, interesting properties, but we will not consider them here.

III

The Golden Square

C. A Study of Length, Area, and Volume

A series of golden squares with a common center, and with the ratio of the lengths of the larger to the smaller of the parallel sides of consecutive golden squares, being equal to the golden ratio, is shown in Figure 52.

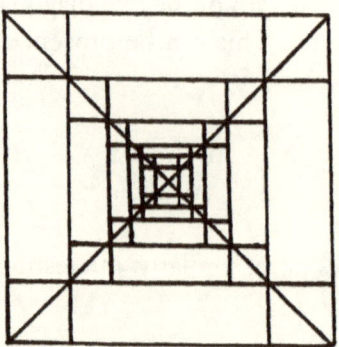

Figure 52

This series of golden squares consists of an infinite number of golden squares, which become, infinitely, dense at the center.

All the vertices of these golden squares lie along the two diagonals of the largest golden square. These vertices divide each of these diagonals into a series of line segments. The ratio of the lengths of the larger to the smaller of consecutive line segments is equal to the golden ratio.

A spiral of line segments, which is based on a series of golden squares with a common center, and with the ratio of the lengths of the larger to the smaller of the parallel sides of consecutive golden squares, being equal to the golden ratio, is shown in Figure 53.

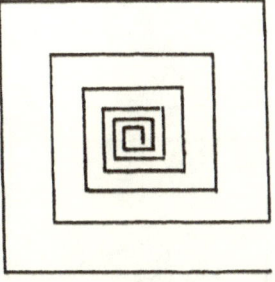

Figure 53

A series of golden squares with a common center, and with the ratio of the areas of the larger to the smaller of consecutive golden squares, being equal to the golden ratio, is shown in Figure 54.

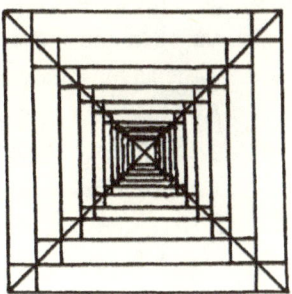

Figure 54

This series of golden squares consists of an infinite number of golden squares, which become, infinitely, dense at the center.

All the vertices of these golden squares lie along the two diagonals of the largest golden square. These vertices divide each of these diagonals into a series of line segments. The ratio of the lengths of the larger to the smaller of consecutive line segments is equal to $\tau^{1/2}$.

This series has two important properties. One property is that the ratio of the area of each square to the area between the square and the next larger, and consecutive square is equal to the golden ratio. This can be proven, as follows. Consider the diagram in Figure 55.

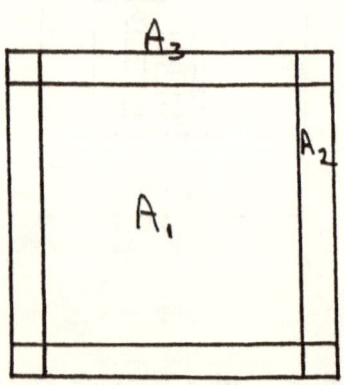

Figure 55

If A_1 = area of the smaller square,
A_2 = area between the smaller, and the larger square, and
A_3 = area of the larger square, then

$$A_3/A_1 = \tau, \text{ but}$$
$$A_3 = A_1 + A_2$$
$$\therefore \ (A_1 + A_2)/A_1 = \tau$$
$$A_1/A_1 + A_2/A_1 =$$
$$1 + A_2/A_1 =$$
$$\therefore \ A_2/A_1 = \tau - 1$$
$$= 1/\tau$$
$$\therefore \ A_1/A_2 = \tau$$

This property holds true for every pair of consecutive golden squares in this series.

The other property is that the ratio of the larger to the smaller of contiguous areas between successive golden squares is equal to the golden ratio. This can be proven, as follows. Consider the diagram in Figure 56.

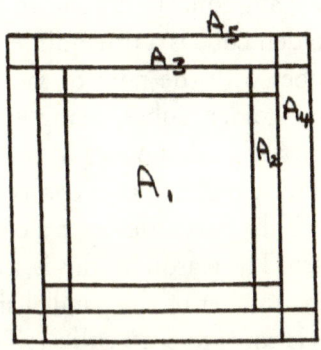

Figure 56

If A_1 = area of the smallest square,

A_2 = area between the smallest, and the middle square,

A_3 = area of middle square,

A_4 = area between the middle, and the largest square, and

A_5 = area of the largest square, then

$$A_1/A_2 = \tau, \text{ and}$$
$$A_3/A_4 = \tau, \text{ but}$$
$$A_3 = A_1 + A_2$$
$$\therefore \ (A_1 + A_2)/A_4 = \tau$$
$$(A_1 + A_2) = \tau * A_4$$
$$(A_1 + A_2)/A_2 = \tau * (A_4/A_2)$$
$$A_1/A_2 + A_2/A_2 =$$
$$\therefore \ \tau + A_2/A_2 =$$
$$\tau + 1 =$$
$$\therefore \ \tau/\tau + 1/\tau = A_4/A_2$$
$$1 + 1/\tau =$$
$$\therefore \ 1 + (\tau - 1) =$$
$$\tau = \qquad , \text{ or}$$
$$A_4/A_2 = \tau$$

This property holds true for every pair of contiguous areas between successive golden squares in this series.

These properties hold true for a series of golden cubes with a common center, and with the ratio of the volumes of the larger to the smaller of consecutive golden cubes, being equal to the golden ratio, except with volume substituted for area. A golden cube is a three-dimensional golden square.

A series of golden cubes with the ratio of the volumes of the larger to the smaller of consecutive golden cubes, being equal to the golden ratio, completely, overlaps a series of golden cubes with the ratio of the lengths of the larger to the smaller of the parallel sides of consecutive golden cubes, being equal to the golden ratio, if the two series have a common center, and if the two series are superimposed. The reason for this is, as follows. The ratio of the lengths of the larger to the smaller of the parallel sides of successive golden cubes in the former series is equal to τ: $\tau^{2/3}$: $\tau^{1/3}$: 1. The ratio of the lengths of the larger to the smaller of the parallel sides of successive golden cubes in the latter series is equal to τ: 1. A side of respective length τ will, always, overlap another side of respective length τ. A side of respective length 1 will, always, overlap another side of respective length 1.

A series of golden cubes with the ratio of the volumes of the larger to the smaller of consecutive golden cubes, being equal to the golden ratio, only overlaps every other golden cube in a series of golden cubes with the ratio of the areas of the larger to the smaller of the parallel faces of consecutive golden cubes, being equal to the golden ratio, if the two series have a common center, and if the two series are superimposed. The reason for this is, as follows. The ratio of the lengths of the larger to the smaller of the parallel sides of successive golden cubes in the former series is equal to τ: $\tau^{2/3}$: $\tau^{1/3}$: 1. The ratio of the lengths of the larger to the smaller of the parallel sides of successive golden cubes in the latter series is equal to τ: $\tau^{1/2}$: 1. A side of respective length τ will, always, overlap another side of respective length τ. A side of respective length 1 will, always, overlap another side of respective length 1. However, a side of respective length $\tau^{1/2}$ will never overlap a side of respective length τ, $\tau^{2/3}$, $\tau^{1/3}$, or 1.

III

The Golden Square

D. A Series of Golden Square Hypercubes, and the Fourth Dimension

By analogy, some of the characteristics of a four dimensional body can be determined. If we move a point through the first dimension we get a line; if we move a line segment through the second dimension, we get polygon which, for clearness, we will assume to be a square; if we move the square through the third dimension we get a cube (or parallelopiped of some sort); therefore, if we move the cube through the fourth dimension, we will get a four dimensional figure. It can be shown, by elaborating this analogy that such a figure has eight cubes, sixteen points, twenty-four planes, and thirty-two lines. (Altieri, 1925, p. 494)

Some mathematicians have tried to show that time is the fourth co-ordinate. If time is the fourth dimension, what would one mean by the fifth dimension or the hundredth dimension? This view is too anthropocentric to be plausible. Time is an attribute of all space. Even in a world of one dimension there would be duration or the time it takes an object to move from one position to another. This view arose from the theory that our space is curved in the fourth dimension. We must resort to analogy to understand what this would mean. Imagine a soap bubble with a cone being pushed through it. A being living on the two dimensional surface of the bubble and having no comprehension of a third dimension would see only the successive cross sections of the cone, which he would notice were continually changing, and he would probably call it aging or growth just as we do here. He would have no suspicion that what he is observing is the motion of a solid object in a third dimension. He would see only the two dimensional cross sections of a three dimensional object. Should a four dimensional object come within the field of our vision, we would see only its three dimensional cross section. Einstein claims that our space is curved in a fourth dimension, and that we are being pushed through from a fourth dimension with a speed equal to that of light, just as the two dimensional surface of a soap bubble is curved in the third dimension and a three dimensional cone can be pushed through it. A three dimensional being could see the whole cone at once, or the whole phenomena occurring which the two dimensional being would interpret as the stream of time. Likewise a four dimensional being could see our whole past, present, and future at once instead of as a stream of time. In any given space, the next higher dimension although spacial has a temporal appearance. Space, time, and motion, are all aspects of relation and some mathematicians claim that they are convertible terms. (Tromp, 1926, pp. 142-143)

In a plane objects as observed by two dimensional beings would have an infinitesimal thickness in a third dimension, otherwise they would be mere shadows, so in our world objects as we observe them would have an infinitismal thickness in a fourth dimension. It is interesting to note that from the discoveries in recent years which seem to indicate that matter moves also in a fourth dimension, the most evidences of this have been found in the infinitesimal, in cellular and molecular activity.

In our experience we find the phenomena of symmetry only in the minute, such as is produced in plants and animals and in corpuscular and atomic action, and not in the large masses, such as mountains. You all know that symmetrical figures can be made to coincide by giving each a quarter turn toward each other through the next higher dimension. From this point of view, symmetrical figures may be regarded as resulting from a splitting of one figure in a given space and an unfolding in the next lower space. This would explain many of the phenomena of corpuscular action. (Ibid., pp. 143-144)

A series of golden square hypercubes can be constructed with the ratio of the lengths of the larger side to the smaller side of consecutive hypercubes, being equal to the golden ratio. See Figure 57.

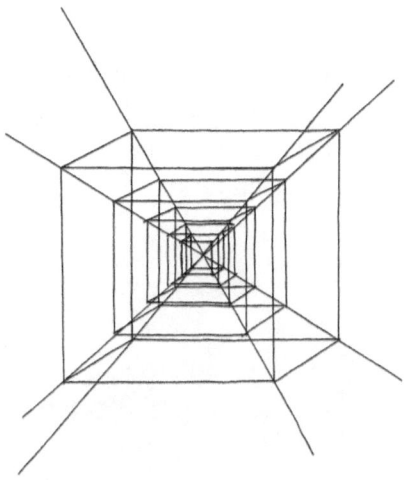

Figure 57

This series consists of an infinite number of golden square hypercubes, which become, infinitely, small at the center.

There are four diagonals, which are, infinitely, long, which intercept the sixteen vertices of each hypercube. The ratio of the lengths of the larger to the smaller of consecutive segments of these diagonals between consecutive golden square hypercubes will be equal to the golden ratio.

The isosceles triangles formed by the four diagonals are, approximately, 54° 40', 54° 40', 70° 40' triangles. Since each isosceles triangle is similar, and since the ratio of the lengths of the larger side to the smaller side of consecutive golden square hypercubes is equal to the golden ratio, then the ratio of the lengths of the two equal larger sides to the two equal smaller sides of consecutive isosceles triangles will be equal to the golden ratio.

The ratio of the volumes of the larger to the smaller of consecutive golden square hypercubes will be equal to τ^3. (Volume=Side*Side*Side) The ratio of the areas of the faces of the larger to the smaller of consecutive golden square hypercubes will be equal to τ^2. (Area=Side*Side)

If the length of the sides of a cube are one, the diagonal of the faces will be equal to $\sqrt{2}$, $(\sqrt{(1^2+1^2)})$. The diagonal of the cube, going through two opposite corners, and the center, will be equal to $\sqrt{3}$, $(\sqrt{(\sqrt{2}^2+1^2)})$.

In the squaring of the circle figure, if the length of the sides of the external square is one, then the relative length of the diagonal of this square will be $\sqrt{2}$. The relative length of the two lateral sides of the inscribed isosceles triangle will be $\sqrt{5}/2$. See Chapter VI., Section A., "The Root Rectangles". Recall, the golden ratio is equal to $(1+\sqrt{5})/2$. If this figure is extended to the third dimension, we will have a cube with inscribed pyramid, and inscribed sphere. If this figure is extended to the fourth dimension, we will have a series of cubes with inscribed pyramids, and inscribed spheres, the ratios of whose dimensions can be equal to some multiple, power, or combination of multiples and powers of the golden ratio.

CHAPTER IV.

The Golden Rectangle

IV

The Golden Rectangle

A. Introduction

A golden rectangle is a rectangle with the ratio of the length of the two longer sides to the length of the two shorter sides, being equal to the golden ratio. See Figure 58.

Figure 58

The geometry of the golden rectangle is fundamental to the geometry of the golden section. It plays an important role in the geometry of the golden spiral of the golden rectangle, the "pseudo-spiral of Fibonaccian growth", the geometric breakdown of the golden rectangle into further golden rectangles, and golden squares, and the regular icosahedron. Three golden rectangles, which lie perpendicular to, and bisect one another, can be formed from the twelve vertices of a regular icosahedron.

Also, the golden rectangle has aesthetic values of its own, which have been, and can be used in art (including drawing, painting, pottery, and sculpture), architecture ("Examples include the Court of the Lions, the Parthenon, and the Cathedral of Notre Dame" (Herz, 1982, p. 52)), and materials' design. In an analysis of the proportions of the bones of the human body, and of other animals, including the horse, the golden rectangle, and other rectangles, whose dimensions are based on the golden section, can be found.

The Greek vase, and other forms of Grecian art are so beautiful not only because they use the golden section, but, also, because they use the themes of the root-two and the root-five rectangles in their overall designs. "This root-five rectangle is the basic shape of vegetable and animal architecture and is the form which has solved the mystery of the perfection of classical Greek art." (Hambidge, 1919, p. 15) The root-five rectangle consists of two golden rectangles, which overlap each other to the extent of a square.

> Both Pythagoras and Euclid called it the rectangle of the Divine Section. In 1876 German psychologist Gustav Theodor Fechner measured the dimensions of thousands of common rectangles, such as playing cards, windows, writing pads, and book covers. He found that, on the average, their proportions were close to the golden mean. Without knowing why, the designers had realized that the golden rectangle was a pleasant shape. Fechner and a successor, Wilhelm Max Wundt, tested hundreds of individuals to determine their preferences for rectangles of various proportions. About 75 per cent preferred the proportions of the golden mean. (Hoffer, 1977, pp. 106-7)

The golden rectangle was, probably, one of the first discovered, and used of all the geometric figures, whose dimensions are based on the golden section. The earliest known evidence of the golden section comes from Egypt.

> The quantity ϕ (often approximated by 10/6 or 16/10) occurs often in the proportions of ancient Egyptian structures erected some 5000 years ago. The "Golden Chamber" of the tomb of Rameses IV measured 16 ells x 16 ells x 10 ells, which is approximately in the ratio of $\phi : \phi : 1$.

Many pieces of Egyptian furniture found in tomb chambers had an overall shape of $\phi : 1 : 1$.

Another volume often used in Egyptian tombs is $\phi^2 : \phi : 1$. The center of a chamber with these dimensions is a distance of ϕ from each of the eight vertices.(West, Griesbach, Taylor, and Taylor, 1982, p. 229)

All of these right-angled parallelepipeds contain the shape of the golden rectangle within them.

Having discovered the golden section, especially, as it is found in the rectangles formed by the proportions of the bones of the human body, the Egyptians, probably, noticed the occurrence of the golden section in other forms of nature, including in trees, shrubs, and flowers. Having noticed the property of pentadactylism, and the property of having five appendages on the trunk in man and the animals, they, probably, also, noticed the occurrence of the number five in the phyllotaxis of the natural flora, that surrounded them. Pentadactylism is the property of "having five fingers or toes on each hand or foot." (*Webster's New World Dictionary of the American Language, College Edition*, 1966, s.v. "pentadactyl")

Having noticed the common reoccurrence of the number five in the environment, that surrounded them in their daily lives, and of the other Fibonacci numbers, especially, the lower ones, including 1, 1, 2, 3, 5, 8, 13, and 21, they discovered the Fibonacci number sequence, and its relationship to the golden section. Having no other way to find the golden ratio, they, probably, found that the best way to estimate the value of the golden ratio was to use the ratio of two of the lower Fibonacci numbers, especially, 5/3 and 8/5. Notice, both of these fractions of consecutive Fibonacci numbers contain the number five. 3/2 and 2/1 would be too broad approximations for the value of the golden ratio.

As mentioned in the quote from West, Griesbach, Taylor, and Taylor, the Egyptians often used the ratios of 10/6 and 16/10 to design the dimensions of their architectural plans. 10/6 reduces to 5/3, and 16/10 reduces to 8/5.

Perhaps the earliest evidence of human knowledge of the golden mean is still visible today on the plateau of Giza in Egypt. Some scholars say that the Egyptians based their work on a loose proportion of 5 to 8, two early numbers of the Fibonacci sequence that approximate the golden mean, but are slightly off. The average ratio of altitude to base of all the pyramids at Giza is about 0.625, or 5 to 8. There is scholarly disagreement as to the exactness of Egyptian knowledge. Further uncertainty is caused by the fact that age has crumbled the capstones of the pyramids so that their precise original heights must be estimated. At least one serious scholar claims that the Great Pyramid was originally 484 feet 5 inches in height. Reduced to inches—there is reason to believe the Egyptians worked in inches—the height would then be 5,813 inches (5, 8, and 13 are numbers in the Fibonacci sequence). (Hoffer, 1977, p. 106)

Other rectangles, whose dimensions are based on the golden section, also, exist, but are not important enough for inclusion here. These other rectangles can be constructed by replacing the sides of a golden rectangle by sides, the ratio of whose lengths is equal to either a multiple, a power, or a combination of multiples and powers of the golden ratio. The purpose of the present study is to elaborate on various geometrical constructions, which depend on, or involve the concept of a golden rectangle.

IV

The Golden Rectangle

B. The Golden Spiral of the Golden Rectangle

The golden spiral of the golden rectangle is a logarithmic spiral. It is one topic associated with the golden rectangle, that other authors, commonly, discuss. The golden spiral of the golden rectangle is important, because it is found, commonly, in nature, and is, occasionally, applied to art. It is found in the curve of a ram's horn, a parrot's beak, an elephant's tusk, and a lion's claw. It is, also, found in the fang of a saber-toothed tiger, and the tusk of a mammoth, both of which are, now, extinct. It is found in the spiral arms of stars of galaxies, the pearly shell of the chambered nautilus, and the human ear.

The golden spiral of the golden rectangle is based on a constant breakdown of the golden rectangle into a series of squares, and golden rectangles, that, eventually, form a continuous spiral. The golden rectangle is unique in that the gnomon, or area left over after the reciprocal rectangle is drawn in, is a square.

By drawing a line segment inside a golden rectangle, parallel to the two shorter sides, and dividing the two longer sides into golden section, a square will be formed on one side, and another golden rectangle will be formed on the other side of this line segment. This can be proven, as follows. Consider the diagram in Figure 59.

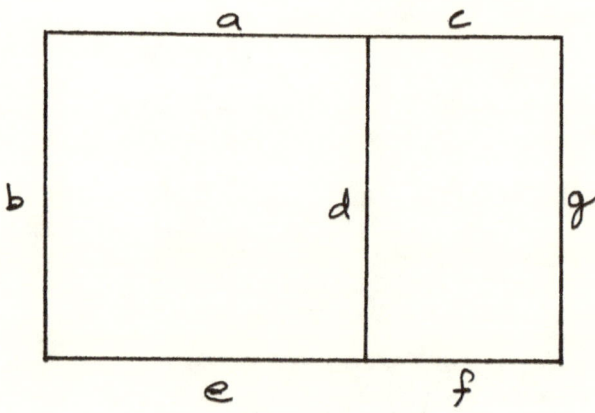

Figure 59

$$a/c = \tau$$
$$\therefore \quad (a+c)/a = \tau$$
$$(a+c)/b = \tau$$
$$\therefore \quad a = b$$
$$\text{But, } b = d = e,$$

so □ abde is a square.

$$\text{Also, } d/c = \tau, \text{ and}$$
$$c = f, \text{ and } d = g.$$

Therefore, □ cdfg is another golden rectangle.

By dividing each smaller and consecutive golden rectangle into a square, and another golden rectangle, a series of squares can be formed, which form a spiral. See Figure 60. It is for this reason that Hambidge called the golden rectangle the "rectangle of whirling squares".

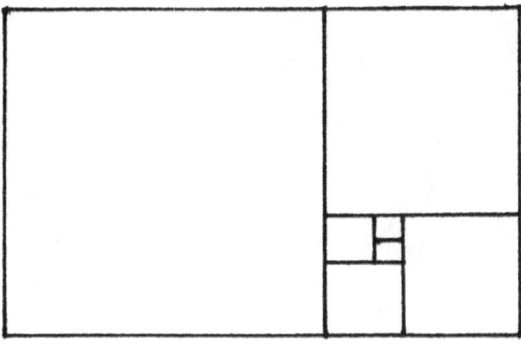

Figure 60

The spiral pole can be determined by finding the point of intersection of corresponding diagonals of two consecutive golden rectangles. See Figure 61.

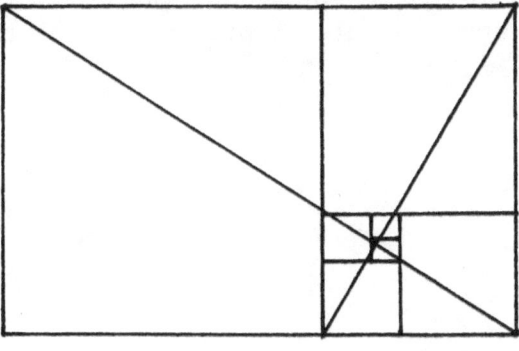

Figure 61

The spiral may evolve around either of four symmetrical spiral poles. See Figure 62.

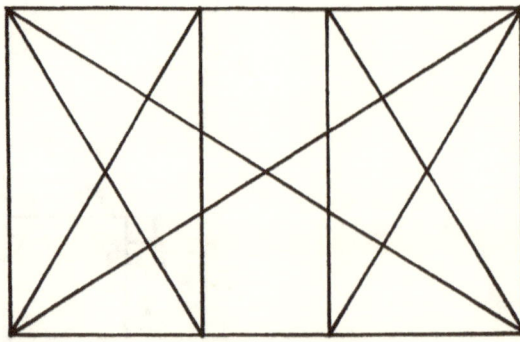

Figure 62

Within the spiral of "whirling squares", a spiral of quadrants of circles, or quarter circles can be formed. See Figure 63. This spiral is called a composite spiral, because it is composed of an aggregate of similar, but different sized parts, which fit together, and, eventually, form a spiral. This composite spiral approaches the curve of the true logarithmic spiral, that goes through the center of each square.

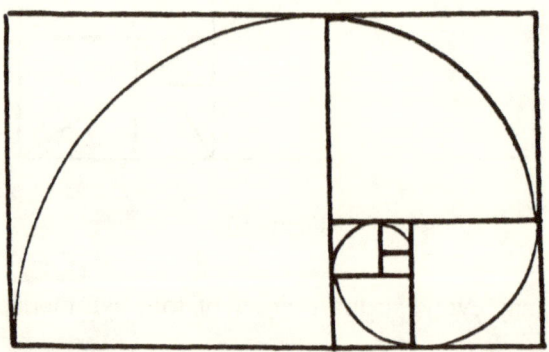

Figure 63

Half of the vertices of all the golden rectangles in the golden spiral of the golden rectangle lie along the two corresponding diagonals of the two largest golden rectangles. See Figure 64. The other half of the vertices of all the golden rectangles lie along the two diagonals, that are drawn between the opposite vertices of alternate golden rectangles. See Figure 64.

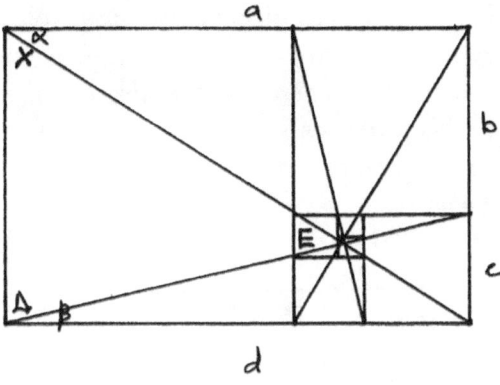

Fig. 64

The measure of the angle between the consecutive radii of this spiral is 45°. This can be proven, as follows. Consider the diagram in Figure 64.

$$\tan(\alpha+\beta) = (\tan\alpha+\tan\beta)/(1-\tan\alpha\tan\beta)$$
$$\tan\alpha = (b+c)/a = 1/\tau, \text{ or } \tau^{-1}$$
$$\tan\beta = c/d = 1/\tau^3, \text{ or } \tau^{-3}$$
$$\therefore \quad \tan(\alpha+\beta) = (\tau^{-1}+\tau^{-3})/(1-\tau^{-1}*\tau^{-3})$$
$$= ((\tau^2+1)/\tau^3)/(1-\tau^{-4})$$
$$= ((\tau^2+1)/\tau^3)/((\tau^4-1)/\tau^4)$$
$$= ((\tau^2+1)/\tau^3)/(\tau^2+1)(\tau^2-1)/\tau^4$$
$$= \tau/(\tau^2-1), \text{ but}$$
$$\tau^2-1 = \tau$$
$$\therefore \quad \tan(\alpha+\beta) = \tau/\tau = 1$$
$$\therefore \quad \alpha+\beta = 45°$$
$$\text{But, } \alpha+X = 90°, \text{ and}$$
$$\Delta+\beta = 90°$$
$$\therefore \quad X = 90°-\alpha, \text{ and}$$
$$\Delta = 90°-\beta$$
$$\text{Also, } \Delta+E+X = 180°$$
$$\therefore \quad (90°-\beta)+E+(90°-\alpha) = 180°, \text{ or}$$
$$180°-\beta-\alpha+E = 180°$$
$$\therefore \quad -\beta-\alpha+E = 0°, \text{ or}$$
$$-(\alpha+\beta)+E = 0°, \text{ or}$$
$$-45°+E = 0°, \text{ or}$$
$$E = 45°$$

Also, corresponding diagonals within consecutive golden rectangles, and corresponding diagonals between pairs of alternate golden rectangles lie at an angle of 90° from each other. Therefore, the measure of the angle between consecutive radii of this spiral is 45°.

A true logarithmic spiral can be constructed, going through the center of each square. See Figure 65.

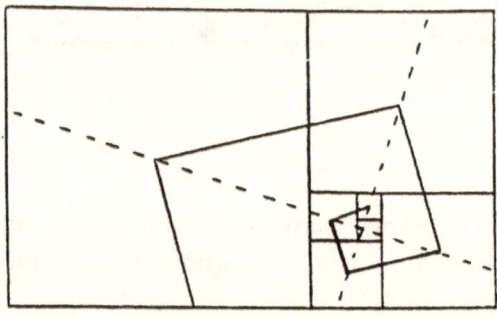

Figure 65

The measure of the internal angle between the radius of the spiral, going to the center of each square in the vertical series of golden rectangles in this spiral, and the horizontal is equal to, approximately, 71° 30'. This can be proven, as follows. Consider the diagram in Figure 66.

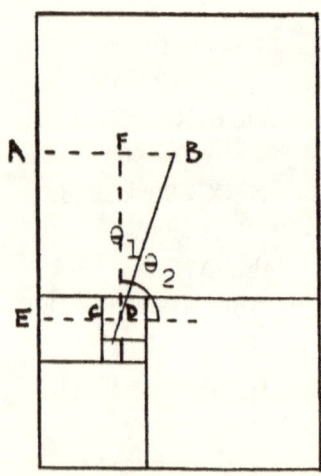

Figure 66

If the length of the sides of the fifth smallest square is one, then, respectively,

$$\overline{CD} = \frac{1}{2},$$

$$\overline{EC} = \tau,$$

$$\overline{AB} = (\frac{1}{2})\tau^4$$

$$\therefore \quad \overline{FB} = \overline{AB} - \overline{CD} - \overline{EC}$$

$$= (\frac{1}{2})\tau^4 - \frac{1}{2} - \tau, \text{ and}$$

$$\overline{FD} = (\frac{1}{2})\tau^4 + \frac{1}{2}.$$

$$\text{Tan}\theta_1 = \overline{FB}/\overline{FD}$$

$$= \frac{(\frac{1}{2})\tau^4 - \frac{1}{2} - \tau}{(\frac{1}{2})\tau^4 + \frac{1}{2}}$$

$$= \frac{\tau^4 - 1 - 2\tau}{\tau^4 + 1}$$

$$= 0.3333332\ldots, \text{ or } 1/3.$$

Notice that $\overline{FB}$ is, exactly, one-third of the length of $\overline{FD}$.

$$\Rightarrow \quad \theta_1 \cong 18° \, 30'$$

Angle BFD is a right angle, or 90°.

$$\overline{FD} \perp \overline{ED}$$
$$\therefore \quad 90° - \theta_1 = \theta_2$$
$$\Rightarrow \quad \theta_2 \cong 71° \, 30'.$$

Notice, the measure of this angle is very close to the measure of an internal angle at the base of a golden triangle, or 72°.

Likewise, the measure of the internal angle between the radius of the spiral, going to the center of each square in the horizontal series of golden rectangles in this spiral, and the vertical is equal to, approximately, 71° 30'.

The true logarithmic spiral of the golden rectangle consists of an infinite number of 31° 40', 58° 20', 90° triangles, rotated around a spiral pole. This can be proven, as follows. Consider the diagram in Figure 67.

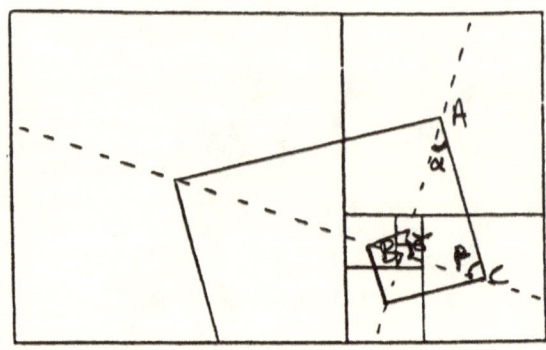

Figure 67

$$\delta = 90°$$
$$\mathrm{Tan}\beta = \overline{AB}/\overline{BC} = \tau$$
$$= 1.6180339\ldots$$
$$\Rightarrow \quad \beta \cong 58° \ 20'$$
$$\Rightarrow \quad \alpha = 90° - \beta \cong 90° - 58° \ 20'$$
$$= 31° \ 40'$$

These triangles are not quite the 30°, 60°, 90° triangles of the one-half of an equilateral triangle. These are the same angles, as the angles at which the diagonal of a golden rectangle lie from the sides of the golden rectangle (see Chapter IV., Section E.).

The true logarithmic spiral of the golden rectangle is completed when a continuous curve is drawn, connecting the centers of the successive squares. See Figure 68.

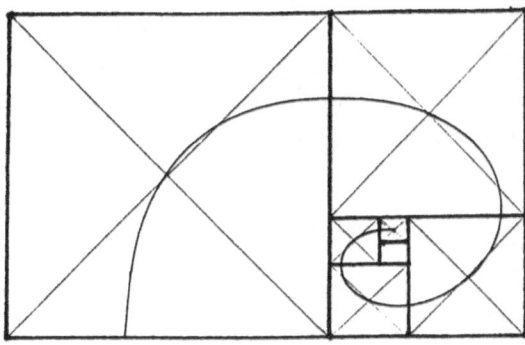

Figure 68

The ratio of the lengths of the longer to the shorter of consecutive chords in this spiral is equal to the golden ratio. This can be proven, as follows. Consider the diagram in Figure 65. The line segments connecting the spiral pole to the centers of successive squares, or radii within successive golden rectangles are similar. Because the ratio of the lengths of the longer to the shorter of corresponding sides of consecutive golden rectangles is equal to the golden ratio, the ratio of the lengths of the longer to the shorter of similar line segments within consecutive golden rectangles is, also, equal to the golden ratio. Therefore, the ratio of the lengths of the longer to the shorter of corresponding radii within consecutive golden rectangles is, also, equal to the golden ratio.

All triangles, formed by two consecutive radii, and one chord of this spiral are similar. The ratio of the lengths of the longer to the shorter of corresponding sides of these consecutive triangles is equal to the golden ratio. Therefore, the ratio of the lengths of the longer to the shorter of consecutive chords within this spiral is equal to the golden ratio. Also, the ratio of the lengths of the longer to the shorter of consecutive arcs in this spiral is equal to the golden ratio. It is for this reason, that this spiral is known as the golden spiral of the golden rectangle.

This spiral can, also, be thought of as an infinite series of right triangles, whose sides are formed by two consecutive radii, and one chord of this spiral. As mentioned before, all these triangles are similar. Therefore, this spiral can, also, be thought of as an infinite series of similar triangles, which grow, successively, larger, or, successively, smaller, as the spiral increases, or decreases, respectively. That is, the shapes of the successive triangles remain the same, only their size changes. Therefore, this spiral can be considered an equiangular, or logarithmic spiral.

The formula for a logarithmic spiral is $r = a * e^{\theta * \cot \alpha}$. "The curve cuts all radii vectors at a *constant angle* α. ($r/r' = \tan\alpha$)." (Yates, 1974, p. 206) "The angle of the spiral can be shown to be the root of the equation $\exp(\frac{1}{2} * \pi \cot\phi) = \tau$, i.e. 73° very nearly." (Cundy & Rollett, 1957, p. 63)

IV

The Golden Rectangle

C. The Geometric Breakdown of the Golden Rectangle into further Golden Rectangles and Golden Squares

The golden rectangle can be broken down, symmetrically, into an infinite number of smaller golden rectangles, and golden squares. This process is demonstrated in Figures 69 through 71.

We have already seen that the golden rectangle can be broken down into a series of squares and further golden rectangles, or a spiral of "whirling squares" (see Section IV., B.). The second step can be accomplished, as follows. By placing the square, as shown in Figure 49 in Section III., B., inside the square on the left-hand side of the golden rectangle, as shown in Figure 59 in Section IV., B., the golden rectangle, shown in Figure 69, will result. It is broken down into five symmetrical parts: two golden rectangles on the left and right-hand sides, two squares on the top and bottom parts of the center section, and a golden rectangle in the center.

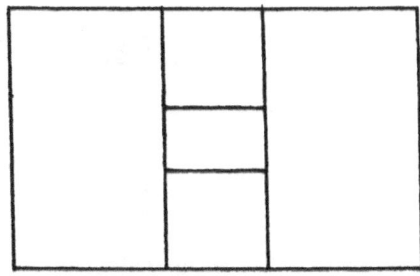

Figure 69

The third step can be accomplished, as follows. By breaking down each golden rectangle in the golden rectangle shown in Figure 69, symmetrically, into further golden rectangles, and each square into a golden square, then the golden rectangle in Figure 70 will result.

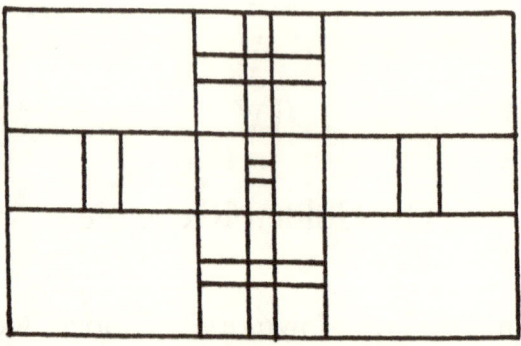

Figure 70

This process can continue, indefinitely. By breaking down each golden rectangle, and square in each previous breakdown, symmetrically, into further golden rectangles, and golden squares, then the original golden rectangle will be broken down, symmetrically, into an infinite number of golden rectangles, and golden squares. A rough approximation of this is shown in Figure 71.

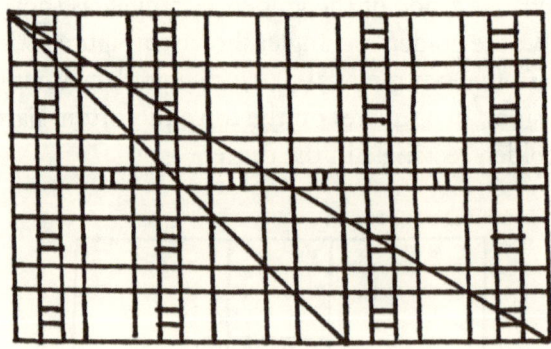

Figure 71

Because the squares are concentric in each golden square, the diagonals of the internal square are segments of the diagonals of the external square.

Diagonals of squares only form diagonals of other squares. See Figure 71. Also, because the golden rectangles are concentric in each golden rectangle pair, the diagonals of the internal golden rectangle are segments of the diagonals of the external golden rectangle. Diagonals of golden rectangles only form diagonals of other golden rectangles. See Figure 71.

IV

The Golden Rectangle

D. The "Pseudo-Spiral of Fibonaccian Growth" (Ghyka, 1977, p. 96)

Table 2 presents two series of alternate fractions of consecutive Fibonacci numbers. The ratio of the larger to the smaller of two consecutive Fibonacci numbers approaches the golden ratio, as the Fibonacci number sequence approaches infinite. Therefore, the values of these two series of alternate fractions approach the golden ratio, as the Fibonacci number sequence approaches infinite. However, these two series approach the value of the golden ratio from opposite sides of the golden ratio on the number line.

Table 2

F_{2n}/F_{2n-1}	F_{2n+1}/F_{2n}
$1/1=1$	$2/1=2$
$3/2=1.5$	$5/3=1.\overline{6}$
$8/5=1.6$	$13/8=1.625$
$21/13=1.6153846\ldots$	$34/21=1.6190476\ldots$
$55/34=1.6176470\ldots$	$89/55=1.6\overline{18}$
$\cdot$	$\cdot$
$\cdot$	$\cdot$
$\cdot$	$\cdot$

The pseudo-spiral of Fibonaccian growth is a spiral, which consists of a finite number of squares rotated around a central square, and which grow, successively, larger. See Figure 72. Each square, and the summation of smaller squares, which form the previous spiral, form a rectangle. The dimensions of each rectangle are one pair of consecutive Fibonacci numbers. The dimensions of the first square are one by one. Because the ratio of the larger to the smaller of two consecutive Fibonacci numbers approaches the golden ratio, as the Fibonacci number sequence approaches infinite, the ratio of the dimensions of each rectangle approaches the golden ratio, as the Fibonacci number sequence approaches infinite. Therefore, the dimensions of each rectangle approach the dimensions of a golden rectangle, as this spiral grows infinitely large.

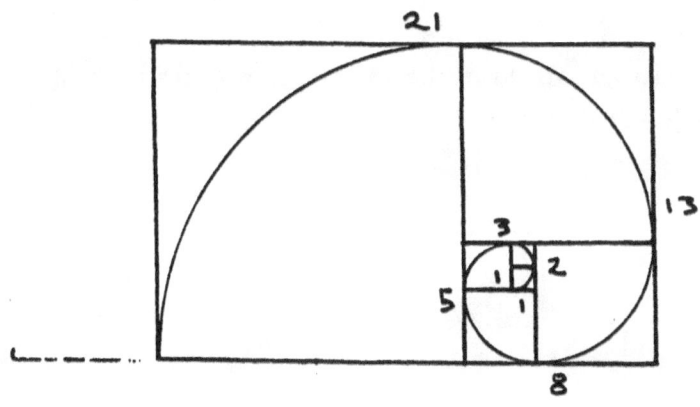

Figure 72

The golden spiral of the golden rectangle is a logarithmic spiral. Because the dimensions of each rectangle approach the dimensions of a golden rectangle, as this spiral grows infinitely large, the pseudo-spiral of Fibonaccian growth approaches the golden spiral of the golden rectangle, as it grows infinitely large. Therefore, the pseudo-spiral of Fibonaccian growth is an approximate logarithmic spiral.

The dimensions of each series of rectangles, either horizontal or vertical, form one of the two series of alternate fractions of consecutive Fibonacci numbers, shown in Table 2. The dimensions of the horizontal series of rectangles in the pseudo-spiral of Fibonaccian growth form the first series of fractions of consecutive Fibonacci numbers, as shown in Figure 73.

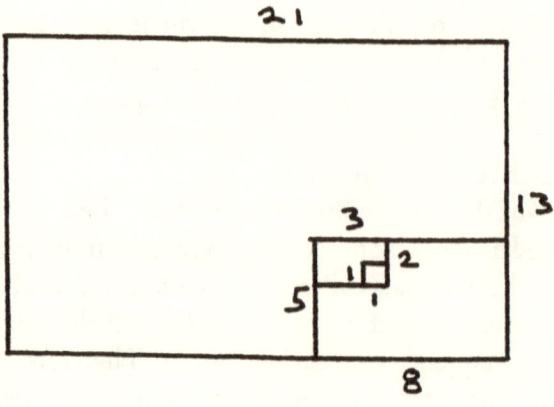

Figure 73

The dimensions of the vertical series of rectangles in the pseudo-spiral of Fibonaccian growth form the second series of fractions of consecutive Fibonacci numbers, as shown in Figure 74.

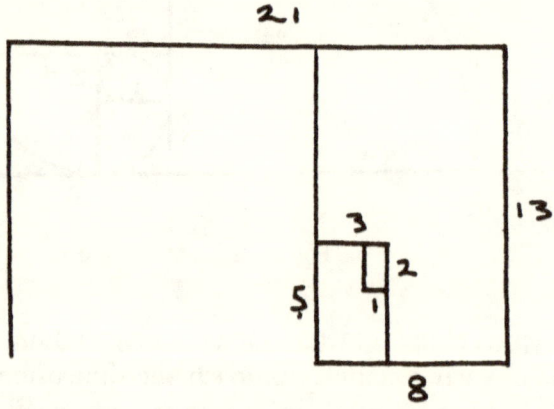

Figure 74

Because the values of the two series of alternate fractions of consecutive Fibonacci numbers approach the value of the golden ratio from opposite sides of the golden ratio on the number line, the shapes of each series of rectangles, either horizontal or vertical, are either more square, or elongated than the shape of the golden rectangle. The shapes of the horizontal series of rectangles are more square than the shape of the golden rectangle. The shapes of the vertical series of rectangles are more elongated than the shape of the golden rectangle.

> Plate XXXII shows the approximative logarithmic spiral connected with a pseudo-gnomonic-cellular growth with Fibonaccian accretion (series 1, 2, 3, 5, 8, 13, 21, 34, 55, 89, . . .); it is often met in biology and is asymptotic to the logarithmic spiral with characteristic ratio and quadrantal pulsation Φ. We have seen that a similar, Fibonaccian, approximation (the progression 55, 89, 144) was probably used in the execution of the Great Pyramid. (Ibid., pp. 94 & 97)

IV

The Golden Rectangle

E. One Golden Rectangle Pair

A golden rectangle pair consists of concentric golden rectangles with the respective sides being parallel. The ratio of the lengths of the parallel sides is equal to either a multiple, a power, or a combination of multiples and powers of the golden ratio.

A golden rectangle pair with the ratio of the lengths of the parallel sides being equal to τ^3 is shown in Figure 75.

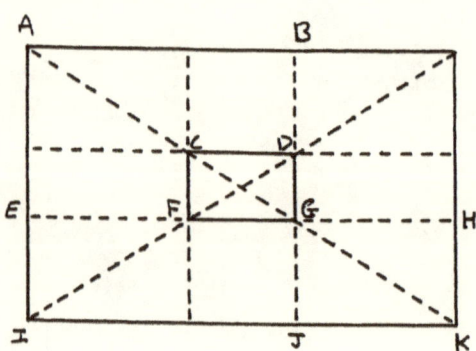

Figure 75

This golden rectangle pair is similar to the golden rectangle diagramed in Figure 69 of "The Geometric Breakdown of the Golden Rectangle into further Golden Rectangles and Golden Squares" (see Section IV., C.).

This golden rectangle pair has some interesting properties. One property is that the ratio of the length of the two line segments, drawn between the two shorter sides, and being parallel to the two longer sides of the external golden rectangle, to the length of the four line segments, drawn from the vertices of the internal golden rectangle to the respective opposite, and shorter sides of the external golden rectangle, is equal to the golden ratio. This can be proven, as follows. Consider the diagram in Figure 75.

$$\overline{IK}/\overline{IJ} = \tau, \text{ but}$$
$$\overline{EH} = \overline{IK}, \text{ and}$$
$$\overline{EG} = \overline{IJ},$$
$$\therefore \quad \overline{EH}/\overline{EG} = \tau$$

Also, the ratio of the length of the two line segments, drawn between the two longer sides, and being parallel to the two shorter sides of the external golden rectangle, to the length of the four line segments, drawn from the vertices of the internal golden rectangle to the respective opposite, and longer sides of the external golden rectangle, is equal to the golden ratio. This can be proven, as follows. Consider the diagram in Figure 75.

$$\overline{AI}/\overline{AE} = \tau, \text{ but}$$
$$\overline{BJ} = \overline{AI}, \text{ and}$$
$$\overline{BG} = \overline{AE},$$
$$\therefore \quad \overline{BJ}/\overline{BG} = \tau$$

A second property is that the ratio of the length of the four line segments, drawn from the vertices of the internal golden rectangle to the respective opposite, and shorter sides of the external golden rectangle, to the length of the four line segments, drawn from the vertices of the internal golden rectangle to the respective shorter sides of the external golden rectangle, is equal to the golden ratio. This can be proven, as follows. Consider the diagram in Figure 75.

$$\overline{EH}/\overline{EG} = \tau$$
$$\therefore \quad \overline{EG}/\overline{GH} = \tau, \text{ but}$$
$$\overline{EF} = \overline{GH},$$
$$\therefore \quad \overline{EG}/\overline{EF} = \tau$$

Also, the ratio of the length of the four line segments, drawn from the vertices of the internal golden rectangle to the respective opposite, and longer sides of the external golden rectangle, to the length of the four line segments, drawn from the vertices of the internal golden rectangle to the respective longer sides of the external golden rectangle, is equal to the golden ratio. This can be proven, as follows. Consider the diagram in Figure 75.

$$\overline{BJ}/\overline{BG} = \tau$$
$$\therefore \ \overline{BG}/\overline{GJ} = \tau, \text{ but}$$
$$\overline{BD} = \overline{GJ},$$
$$\therefore \ \overline{BG}/\overline{BD} = \tau$$

A third property is that the ratio of the length of the four line segments, drawn from the vertices of the internal golden rectangle to the respective shorter sides of the external golden rectangle, to the length of the two longer sides of the internal golden rectangle is equal to the golden ratio. This can be proven, as follows. Consider the diagram in Figure 75.

$$\overline{EG}/\overline{EF} = \tau$$
$$\therefore \ \overline{EF}/\overline{FG} = \tau$$

Also, the ratio of the length of the four line segments, drawn from the vertices of the internal golden rectangle to the respective longer sides of the external golden rectangle, to the length of the two shorter sides of the internal golden rectangle is equal to the golden ratio. This can be proven, as follows. Consider the diagram in Figure 75.

$$\overline{BG}/\overline{BD} = \tau$$
$$\therefore \ \overline{BD}/\overline{DG} = \tau$$

A fourth property is that the ratio of the length of the two diagonals of the external golden rectangle to the length of the four line segments, drawn from the vertices of the internal golden rectangle to the respective opposite vertices of the external golden rectangle, is equal to the golden ratio. This can be proven, as follows. Consider the diagram in Figure 75.

$$\Delta AIK \backsim \Delta AEG$$
$$\overline{IK}/\overline{IJ} = \tau, \text{ but}$$
$$\overline{EG} = \overline{IJ},$$
$$\therefore \ \overline{IK}/\overline{EG} = \tau$$
$$\therefore \ \overline{AK}/\overline{AG} = \tau$$

A fifth property is that the ratio of the length of the four line segments, drawn from the vertices of the internal golden rectangle to the respective opposite vertices of the external golden rectangle, to the length of the four line segments, drawn from the vertices of the internal golden rectangle to the respective vertices of the external golden rectangle, is equal to the golden ratio. This can be proven, as follows. Consider the diagram in Figure 75.

$$\overline{AK}/\overline{AG} = \tau$$
$$\therefore \ \overline{AG}/\overline{GK} = \tau, \text{ but}$$
$$\overline{GK} = \overline{AC},$$
$$\therefore \ \overline{AG}/\overline{AC} = \tau$$

A sixth property is that the ratio of the length of the four line segments, drawn from the vertices of the internal golden rectangle to the respective vertices of the external golden rectangle, to the length of the two diagonals of the internal golden rectangle is equal to the golden ratio. This can be proven, as follows. Consider the diagram in Figure 75.

$$\overline{AG}/\overline{AC} = \tau$$
$$\therefore \ \overline{AC}/\overline{CG} = \tau$$

A seventh property is that the angle formed between one diagonal, and one of the longer sides of a golden rectangle is equal to 31° 44'. This can be proven, as follows. Consider the diagram in Figure 76.

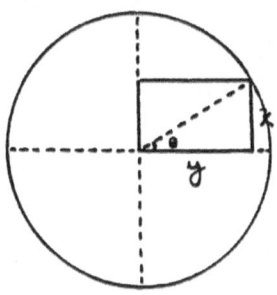

Figure 76

$$\cot\theta = y/x$$
$$= \tau$$
$$= 1.6180\ldots$$
$$\therefore \ \theta \cong 31° 44'$$

CHAPTER V.

The Pentagram

V

The Pentagram

A. Introduction

A pentagram is a five pointed star with five equal sides, and five equal angles. It can be proven that each side divides each other side, that it intersects, into golden ratio (see Section B., Chapter V.). See Figure 77.

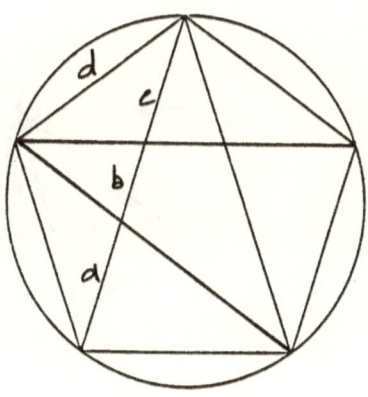

Figure 77

$$(a+b+c)/(a+b) = (a+b)/c$$
$$= d/c$$
$$= a/b$$
$$= \tau$$

A pentagram may be constructed by extending the sides of a regular pentagon, until they touch. See Figure 78.

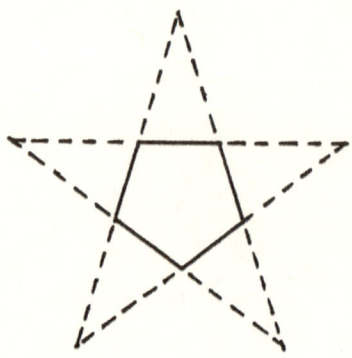

Figure 78

A pentagram may, also, be constructed by inscribing diagonals inside a regular pentagon. See Figure 79.

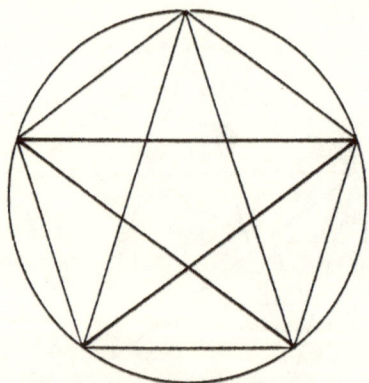

Figure 79

Lastly, a pentagram can be constructed from the basis of a golden triangle, and the two angle bisectors at the base of the golden triangle. A pentagram may be constructed by extending the two angle bisectors at the base of a golden triangle, until they are of an equal length to the length of the two lateral sides of the golden triangle. See Figure 80a. Then, by connecting the two new endpoints of these extensions, a five pointed star will be formed. See Figure 80b. The length of this last line segment will be of an equal length to the length of the two lateral sides of the golden triangle, and to the length of the two angle bisectors at the base of the golden triangle, and their extensions. By removing the base of the golden triangle, the pentagram will be formed. See Figure 80c.

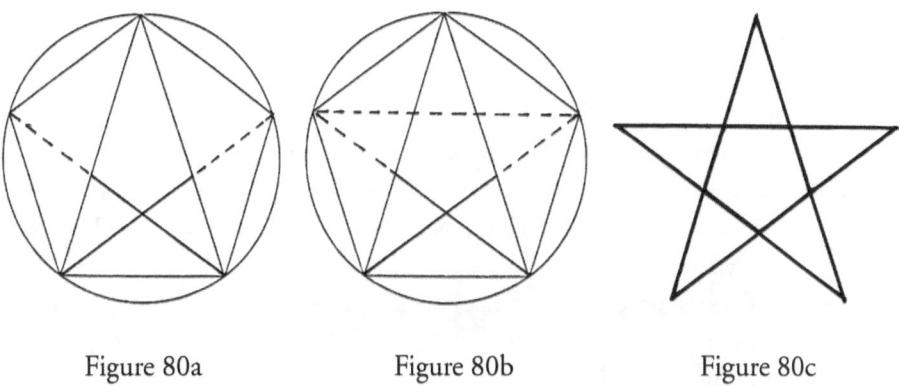

Figure 80a   Figure 80b   Figure 80c

The pentagram has for a long time been used as a symbol of secrecy, or sorcery. This fact is, probably, due to its association with the golden section. The pentagram is thought by many to have originated with the Pythagoreans. But, it is known that the Egyptians, thousands of years, earlier, used a squaring of the circle grid with inscribed cross and diagonals to draw the pentagram.

> Brunés shows how the ancient Egyptians used the basic design of a circle inscribed in a square to divide both circle and square geometrically into equal parts from 2 to 10, and all their possible multiples, without recourse to measuring of arithmetical calculations, with the aid of nothing but a straightedge and a compass—common emblems, along with the Pyramid, of the Masonic orders of yesterday and today. (Tompkins, 1971, p. 261)

> Brunés found that the circle was indeed considered sacred by the Egyptians, as were the square and the cross and the triangle, all of which are intimately incorporated into the Great Pyramid with its square base and triangular faces designed to represent the "sacred" circle.

Brunés demonstrates how the circle inscribed in a square and quartered by a cross enabled the ancient Egyptian geometer to inscribe in a circle the basic figures of pentagon, hexagon, octagon and decagon.

Of these the pentagon with its five-pointed star is perhaps the most important: it automatically produces the Golden Section and the ϕ proportion in the simplest geometric manner.

Each line of the five-pointed star—the symbolic sign of recognition of the initiated Pythagorean, whose hermetic meaning if meant death to reveal—cuts the other in the proportion of major to minor: the Golden Section.

Furthermore, the side of a pentagon inscribed in a circle whose circumference is equal to the perimeter of the Pyramid will be equal to the apothem, or slant height, of the Pyramid, which will be the value of ϕ.

A pentagon divides a circle in 72° segments. With the main cross, the pentagon radii form angles of 18°, 36°, 54° and 72°. (Ibid., pp. 261-262)

Pentacles, or pentagrams, during the Middle Ages, represented money, or the merchant class. Pentacles were, also, known as "shekels".

The pentagram has for a long time been used as the symbol, drawn by children for fun, and by artists for other reasons to represent stars in outer space. This is, probably so, because the pentagram is producible by one continuous line segment. Also, it is the star with the least number of vertices, that can be inscribed in a regular polygon, and made from the chords of that polygon.

According to the *Webster's Third New International Dictionary* (1971), the word pentagram comes from the Greek word pentagrammon. Penta means five. Grammon represents the Greek gramma letter, and means letter of the alphabet.

Other names for the pentagram are pentacle, pentalpha, and pentangle. The word pentacle comes from the Old Italian word pentacol, or pentacolo, which is thought to come from the Medieval Latin word pentaculum. Penta means five. Culum is a Latin diminutive suffix, which means smallness, or decreasing in size. Pentalpha is a Greek word. Penta or pente means five. Alpha is the first letter of the Greek alphabet. According to the *Webster's Third New International Dictionary* (1971), this word was derived "from its presenting the form of an A on each of its five corners" (s.v. "pentalpha"). The word pentangle, simply, means five-angled.

The geometry of the pentagram is fundamental to the geometry of the golden section. It plays an important role in the geometry of the regular pentagon, the regular decagon, the dodecahedron, the icosahedron, and the star-dodecahedron. Also, the pentagram has aesthetic values of its own, which have been, and can be used in art, architecture, and materials' design. For example, the designs of many geodesic domes are based on a regular pentagon shape, and/or a regular hexagon shape. The designs of most geodesic domes are based on the expanded forms of either a dodecahedron, or an icosahedron. The designs of geodesic domes can, also, be based on the expanded forms of any of the thirteen semiregular solids.

Also, the property of pentadactylism is common amongst man and the animals. Animals, that are pentadactyl, include the frog, the salamander, the lizard, the skink, the crocodile, the squirrel, the possum, the koala bear, the raccoon, many vertebrates, and many mammals, including all monkeys and apes. With arms and legs, outstretched, man can obtain the shape of an irregular pentagram. The five senses, located on the front of the faces of all men form the shape of an irregular pentagram.

The purpose of the present study is to elaborate on various geometrical constructions, which depend on, or involve the concept of a pentagram.

V

The Pentagram

B. Two Types of Towers, Made From Pentagrams

In a regular pentagon, if five diagonals can be drawn in, they will be of an equal length, at equal angles from each other, and intersect each other in equivalent segments. The resulting, internal polygon will be a pentagram with the five apices, having angles of 36°. This can be proven, as follows. Consider the diagram in Figure 81.

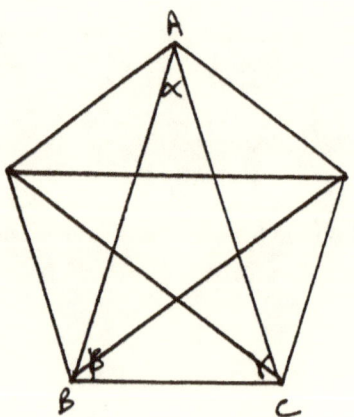

Figure 81

$$\alpha = (1/2)(1/5)*360°$$
$$= (1/10)360°$$
$$= 36°$$

Since triangle ABC is an isosceles triangle, angles β and γ will be equal.

$$\angle\alpha + \angle\beta + \angle\gamma = 180°$$
$$\angle\alpha + 2\angle\beta = 180°$$
$$36° + 2\beta = 180°$$
$$2\beta = 180° - 36°$$
$$= 144°$$
$$\Rightarrow \quad \beta = \gamma = 72°$$

Therefore, triangle ABC is a golden triangle. Therefore,

$$\overline{AB}/\overline{BC} = \overline{AC}/\overline{BC} = \tau$$

All the internal angles of a regular pentagon are equal to 108°. This can be proven, as follows. Consider the diagram in Figure 82.

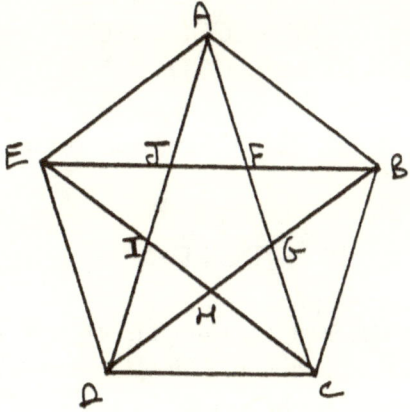

Figure 82

$$\angle BCD = (1/2)(3/5)360°$$
$$= (3/10)360°$$
$$= 3*36°$$
$$= 108°$$

Sides $\overline{BC}$ and $\overline{CD}$ are equal, so $\triangle BCD$ is an isosceles triangle. Therefore, angles CBD and BDC are equal, and can be calculated.

$$\angle CBD + \angle BDC + \angle BCD = 180°$$
$$2\angle BDC + \angle BCD = 180°$$
$$2\angle BDC + 108° = 180°$$
$$2\angle BDC = 72°$$
$$\therefore \quad \angle BDC = \angle CBD = 36°$$

Therefore, ΔBCD is another form of a golden triangle (see Section B., Chapter II.). Therefore,

$$\overline{BD}/\overline{CD} = \overline{BD}/\overline{BC} = \tau$$

Triangles DAC and DGC are similar. This can be proven, as follows. Consider the diagram in Figure 82.

$$\angle BDC = 36°, \text{ but}$$
$$\angle ACD = 72°$$
$$\angle BDC + \angle ACD + \angle CGD = 180°$$
$$36° + 72° + \angle CGD = 180°$$
$$108° + \angle CGD = 180°$$
$$\Rightarrow \quad \angle CGD = 72°$$

Therefore, ΔDGC is a 36°, 72°, 72° triangle, or another golden triangle. Therefore,

$$\overline{DC} = \overline{DG}$$

Each diagonal in this regular pentagon, or each side of the inscribed pentagram divides each other diagonal, or side, respectively, that it intersects, into golden section. This can be proven, as follows. Consider the diagram in Figure 82.

$$\overline{AD}/\overline{DC} = \overline{DC}/\overline{GC}$$
$$\overline{AD}/\overline{DG} = \text{, but}$$
$$\overline{AD} = \overline{DB}$$
$$\therefore \quad \overline{DB}/\overline{DG} = \overline{DG}/\overline{GC}, \text{ but}$$
$$\overline{GC} = \overline{GB}, \text{ so}$$
$$\overline{DB}/\overline{DG} = \overline{DG}/\overline{GB}$$

Therefore, line segment DB is divided into golden section at point G by line segment AC. This holds true for every diagonal in the regular pentagon.

Each diagonal is intersected twice by two other diagonals at two symmetrical points on its length.

The pentagram can be used to construct many different geometric constructions. Two of these are two types of towers, made from pentagrams. The first type of tower, made from pentagrams, is a series of ever diminishing pentagrams, which are constructed, vertically, inside of a pentagram. See Figure 83.

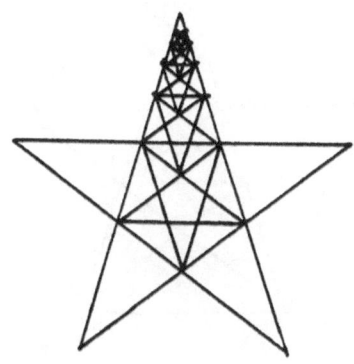

Figure 83

This tower can be started by drawing a regular pentagram, large enough, so that other, smaller, pentagrams can be constructed inside it. After the first pentagram is drawn, a second pentagram can be drawn in the interior pentagon of the first pentagram. This is done, so that each of the five vertices of the second pentagram, also, form one of the five vertices of the regular pentagon on the interior of the first pentagram. See Figure 83. Then, a third pentagram can be drawn on top of the second pentagram, and so on.

Since the two sides of each successive pentagram, that form the bottom apices, lie at an angle of 72° from the horizontal, each pair of these sides are parallel to the two, respective, lateral sides of the external golden triangle. The two sides of each successive pentagram, that form the two upper apices, and are joined by the horizontal side, lie at an angle of 36° from the horizontal. Therefore, these two sides are equivalent to the first two angle bisectors at the base of each successive, and smaller golden triangle in the tower. The horizontal side of each successive pentagram is equivalent to the base of each golden triangle formed, as the tower grows, infinitely, small. Therefore, the third type of tower made from golden triangles, as described in Section D.,

Chapter II., can be superimposed, and, completely, overlaps this tower of pentagrams.

The second type of tower, made from pentagrams, can be constructed by extending the diagonals (struts) of a second type of tower, made from golden triangles, so that they are of an equal length to the two, remaining lateral sides of the external golden triangle. See Figure 84.

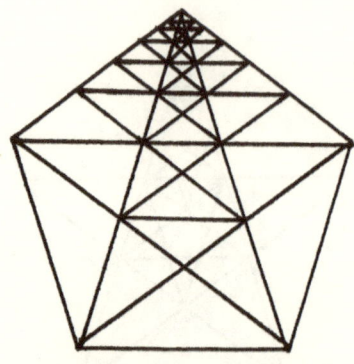

Figure 84

Then, the horizontal side of each respective pentagram will form the line segment of each consecutive, and smaller strut of the third type of tower, made from golden triangles.

This tower can, easily, be constructed inside the pentagon, which circumscribes the external pentagram. This type of tower, made from pentagrams, consists of an infinite number of pentagrams, which grow, successively, smaller, as they reach the apex of the external pentagram.

V

The Pentagram

C. Three Ten-Pointed Stars

Three regular ten-pointed stars can be constructed. The first can be constructed by connecting every second consecutive vertex of a regular decagon in a continuous fashion in either a clockwise, or counterclockwise fashion. See Figure 85. This star may, also, be constructed by inscribing two overlapping, regular pentagons inside a regular decagon.

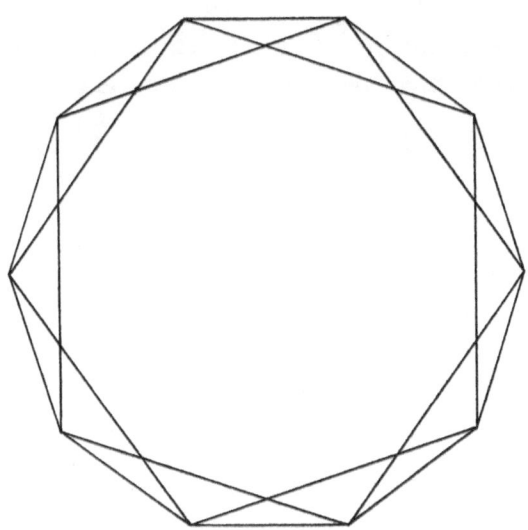

Figure 85

The second ten-pointed star can be constructed by connecting every third consecutive vertex of a regular decagon in a continuous fashion in either a clockwise, or counterclockwise direction. See Figure 86. This star may, also, be constructed by inscribing ten golden triangles inside a regular decagon, so that each respective apex, also, forms one of the vertices of the regular decagon. Then, by drawing in the two altitudes of these golden triangles at the bases of the golden triangles, the ten-pointed star will be formed.

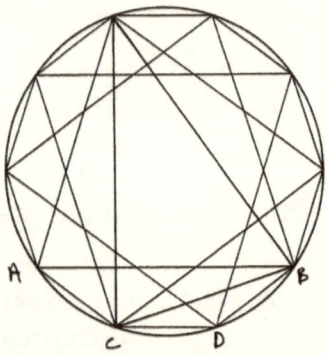

Figure 86

The third ten-pointed star can be constructed by connecting every fourth consecutive vertex of a regular decagon in a continuous fashion in either a clockwise, or counterclockwise direction. See Figure 87. This star may, also, be constructed by inscribing two overlapping pentagrams inside a regular decagon.

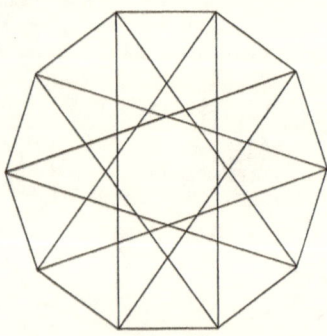

Figure 87

The angle formed between one of the sides of the second ten-pointed star, and the respective adjacent base of the golden triangle, forming an acute angle with the side, cuts out one of the sides of the regular decagon. This side subtends an arc of one-tenth of the perimeter of the circle, that circumscribes the regular decagon. Also, the vertex of an angle, so formed, lies on the perimeter of the circle, that circumscribes the regular decagon. Therefore, the measure of an angle, so formed, is equal to 18°. This can be proven, as follows. Consider the diagram in Figure 86.

$$\angle ABC = (1/2)(1/10)360°$$
$$= (1/20)360°$$
$$= 18°$$

The angle formed between one of the sides of the second ten-pointed star, and the respective adjacent side of the regular decagon, forming an acute angle with the side, cuts out two of the sides of the regular decagon. These two sides subtend an arc of two-tenths of the perimeter of the circle, that circumscribes the regular decagon. Also, the vertex of an angle, so formed, lies on the perimeter of the circle, that circumscribes the regular decagon. Therefore, the measure of an angle, so formed, is equal to 36°. This can be proven, as follows. Consider the diagram in Figure 86.

$$\angle ABD = (1/2)(2/10)360°$$
$$= (2/20)360°$$
$$= (1/10)360°$$
$$= 36°$$

Therefore, the measure of the angle formed between the base of the golden triangle, and the respective adjacent side of the regular decagon, forming an acute angle with the base, is equal to 18°. This can be proven, as follows. Consider the diagram in Figure 86.

$$\angle CBD = \angle ABD - \angle ABC$$
$$= 36° - 18°$$
$$= 18°$$

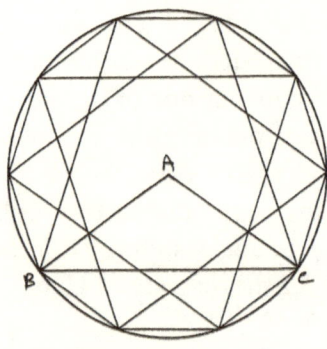

Figure 88

In the second ten-pointed star, the ratio of the length of one of the sides to the length of the radius of the circle, that circumscribes the ten-pointed star, is equal to the golden ratio. This can be proven, as follows. The angle formed by two radii of the circle, that circumscribes the ten-pointed star, spaced three vertices apart, cuts out three of the sides of the regular decagon, that circumscribes the ten-pointed star. These three sides subtend an arc of three-tenths of the perimeter of the circle, that circumscribes the ten-pointed star. Also, the vertex of an angle, so formed, is the center of the circle, that circumscribes the ten-pointed star. Therefore, the measure of an angle, so formed, is equal to 108°. This can be proven, as follows. Consider the diagram in Figure 88.

$$\angle BAC = (3/10)360°$$
$$= 108°$$

All radii of a circle are equivalent. Therefore, all triangles, formed by two radii of the circle, that circumscribes the ten-pointed star, and one of the sides of the ten-pointed star, are isosceles triangles. Therefore, the measures of the two, smaller angles are equal, and can be calculated:

$$\angle ABC = \angle ACB$$
$$\angle ABC + \angle BAC + \angle ACB = 180°$$
$$2\angle ABC + \angle BAC = 180°$$
$$\therefore \quad 2\angle ABC = 180° - \angle BAC$$
$$= 180° - 108°$$
$$= 72°$$
$$\therefore \quad \angle ABC = \angle ACB = 36°$$

Therefore, all triangles, so formed, are 36°, 36°, 108° triangles. The ratio of the length of the longer side to the length of the two shorter sides is equal to the golden ratio (see Section B., Chapter II.). Therefore, the ratio of the length of one of the sides of the ten-pointed star to the length of the radius of the circle, that circumscribes the ten-pointed star, is, also, equal to the golden ratio.

In any golden triangle, inscribed in a regular decagon, the line segment, which goes from one base vertex of the golden triangle to the respective third, consecutive vertex of the regular decagon, and forming an acute angle with the base of the golden triangle is, also, the altitude, and its extension at the base of the golden triangle. This can be proven, as follows. We have, already, found that such a line segment lies at an angle of 18° from the base of the golden triangle. This is the same angle, as the angle at which the altitude at the base, lies from the base of a golden triangle. See Figure 89.

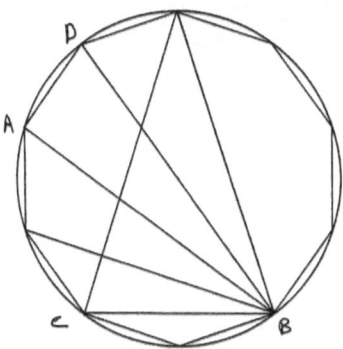

Figure 89

The line segment, which goes from one base vertex of the golden triangle to the respective fourth, consecutive vertex of the regular decagon, and forming an acute angle with the base of the golden triangle is, also, the angle bisector, and its extension at the base of the golden triangle. This can be proven, as follows. Consider the diagram in Figure 89.

$$\angle ABC = (1/2)(2/10)360°$$
$$= (2/20)360°$$
$$= 36°$$

This is the same angle, as the angle at which the angle bisector at the base, lies from the base of a golden triangle.

The line segment, which goes from one base vertex of the golden triangle to the respective fifth, consecutive vertex of the regular decagon is, also, the circumradius, and its extension at the base of the golden triangle. This can be proven, as follows. Consider the diagram in Figure 89.

$$\angle DBC = (1/2)(3/10)360°$$
$$= (3/20)360°$$
$$= 54°$$

This is the same angle, as the angle at which the circumradius at the base of the golden triangle lies from the base of a golden triangle.

The line segment, that goes through the circumcenter of the golden triangle, also, goes through the center of the regular decagon, and the center of the circle, that circumscribes the golden triangle.

V

The Pentagram

D. The Golden Diamond

In two dimensions, a diamond shaped figure can be constructed by drawing two equivalent, isosceles triangles, so that they are inverted, and have the same base. The golden diamond is a pentagonal based diamond, the lengths of whose sides are based on the golden section. It is a three-dimensional structure of my own construction, and can be formed from a regular pentagram. See Figure 90.

Figure 90

The measure of the angle of each apex of a regular pentagram is 36°. Also, each triangle formed by two equal segments of the sides of a pentagram, and one of the sides of the internal, regular pentagon is an isosceles triangle.

Therefore, each such triangle is a golden triangle. By bending down each of these five triangles, so that an upside down pentagonal based pyramid is formed, the bottom part of the golden diamond will be constructed. Then, each lateral face of this upside down, pentagonal based pyramid will be a golden triangle. See Figure 91.

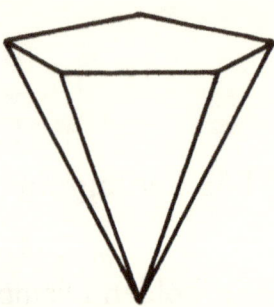

Figure 91

Now, the top part can be constructed. By constructing equilateral triangles on the sides of the pentagonal base of the bottom pyramid, and then folding these equilateral triangles up, so that their apices touch, a pyramid will be formed. This pyramid will have a pentagonal base, and equilateral triangles, as faces. See Figure 92.

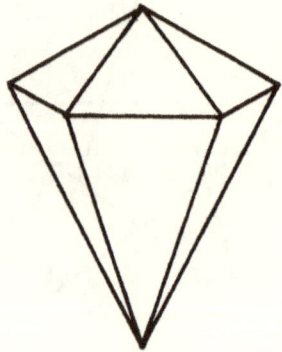

Figure 92

This pentagonal based pyramid can be truncated in either one of two ways, according to the golden section. This pyramid can be cut, so that the ratio of the length of equal segments at the base of the two lateral sides of each equilateral triangle to the length of the truncated side is equal to the golden ratio. See Figure 93. Or, this pyramid can be cut, so that the length of equal segments at the base of the two lateral sides of each equilateral triangle is equal to the length of each truncated side. See Figure 94. Or, this pyramid can be cut, so that the ratio of the length of the truncated side to the length of equal segments at the base of the two lateral sides of each equilateral triangle is equal to the golden ratio. See Figure 95.

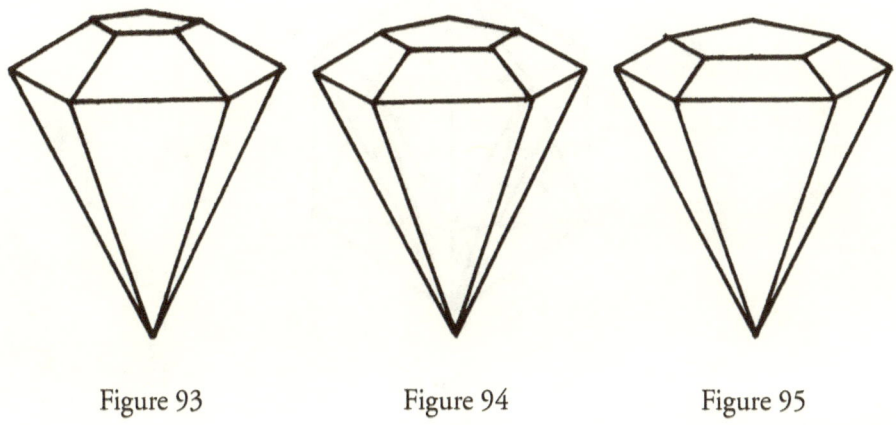

Figure 93   Figure 94   Figure 95

A cube can be inscribed inside a regular dodecahedron, so that each edge of the cube forms one diagonal of one of the faces of the regular dodecahedron. A regular icosahedron can be inscribed inside an octahedron, so that each vertex of the regular icosahedron touches, and divides the respective edge of the octahedron, that it intersects, into golden section. Also, three equivalent golden rectangles can be inscribed inside a regular icosahedron, which bisect each other, and lie in perpendicular planes. See Appendix IV., "The Five Regular Solids".

Now, we can find that a golden diamond, with or without a truncated, top pyramid, can be inscribed inside a regular icosahedron. If the top pyramid is not truncated, we find that it forms, exactly, one of the pentagonal based pyramids, formed from five adjacent faces of the regular icosahedron, which join at a common apex. See Figure 96.

Now, if we draw line segments from each of the five vertices of the pentagonal base to a vertex of the regular icosahedron on the opposite side of the icosahedron, one will find that each of these line segments forms one of the diagonals of the respective pentagonal base, that it transects. All the faces of any of these pentagonal based pyramids have edges of an equal length to the edges of the base of the pyramid. Also, the ratio of the length of any diagonal of a pentagon to the length of any side is equal to the golden ratio. Therefore, the ratio of the length of any diagonal of the pentagonal base of any of the pentagonally based pyramids, that are formed from the faces of a regular icosahedron, to the length of any of the edges of the regular icosahedron is equal to the golden ratio.

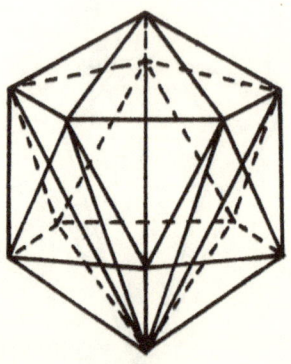

Figure 96

If the top pyramid is truncated, it can still be inscribed inside one of the pentagonally based pyramids, that are formed from the faces of a regular icosahedron, because its faces, also, were, originally, equilateral triangles. Because an icosahedron can be inscribed in a sphere, a golden diamond can be inscribed in a sphere. A golden diamond without a truncated top pyramid has seven vertices, fifteen edges, and ten faces. A golden diamond with a truncated top pyramid has eleven vertices, twenty edges, and eleven faces. A regular icosahedron has twelve vertices, thirty edges, and twenty faces.

In the golden diamond, if one of the diagonals of the pentagon, that forms the base of both pyramids, and the apex of the base pyramid are placed in the plane of the page, then the base pyramid will form an inverted equiangular triangle in the plane of the page.

CHAPTER VI.

The Golden Section in Art and Architecture

VI

The Golden Section in Art and Architecture

A. The Root Rectangles

The Greeks developed the science of the root rectangles, and the theory of irrationals, early in the history of Greek mathematics, and for the remainder of its development, it remained an integral part. This science was, first, adopted from the Egyptians, and had its basis in nature, especially, in plant growth, or phyllotaxis, and in the rectangles formed by the proportions of the bones of the human body. The study of the root rectangles was developed as a science, because of its importance in art, and architecture, and as a means of transferring the fundamental structures of nature to the ideas of art. The reason for a brief description of the root rectangles at this stage of the book is to acquaint the reader with a knowledge of these rectangles in order for him to understand the basic rectangles, that are found, as the design basis of nature, and, also, the working of Greek art and architecture.

Many of the ideas in this section are adopted from the book, *The Diagonal*, by Jay Hambidge (1919), who discusses dynamic symmetry, and who, perhaps, was the first to rediscover the basis of design in Greek art and architecture.

The root two rectangle is one of the simplest, and was, probably, one of the first of the root rectangles discovered by the Greeks. Its design is based on the geometry of a square. It is one of the rectangles used, most commonly, as the overall shape, and as a design basis for cultural artifacts, that have been found, and have survived from the classical age of Greece. It is a rectangle with the ratio of the length of the two longer sides to the length of the two shorter sides, being equal to the square root of two. See Figure 97.

Figure 97

A root two rectangle can be constructed by drawing in the diagonal of a square. Then, extend this diagonal along the base. Then, by filling in the missing sides, a root two rectangle will be constructed. See Figure 98.

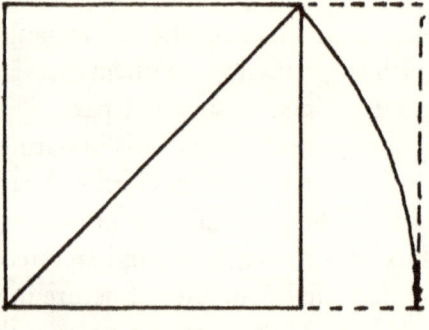

Figure 98

A root two rectangle is constructed from a square, and its diagonal. If the length of the sides of the square are one, then the length of the diagonal of this square will be the square root of two, using the Pythagorean theorem. Therefore, if the diagonal is extended along the base, the ratio of the dimensions of the new rectangle will be $\sqrt{2}:1$, or $\sqrt{2}$.

The root five rectangle is the rectangle, the ratio of the lengths of whose sides was the greatest, used by the Greeks. It is a rectangle, which is found, commonly, in the rectangles formed by the proportions of the bones of the human body. It is the rectangle, which was used most commonly in Greek art and architecture, as the overall shape, and as a design basis for cultural artifacts, that have been found, and have survived from the classical age of Greece. The root five rectangle is a rectangle with the ratio of the length of the two longer sides to the length of the two shorter sides, being equal to the square root of five. See Figure 99.

Figure 99

A root five rectangle can be constructed by drawing in the two diagonals to the midpoint of the base of a square. Then, these two diagonals are extended along the base. Then, by filling in the missing sides, a root five rectangle will be constructed. See Figure 100.

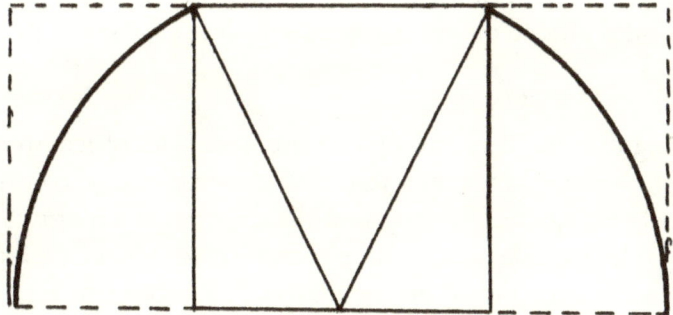

Figure 100

A root five rectangle is constructed from a square, and the diagonals to the midpoint of the base of the square. If the length of the sides of the square is two, then the length of the diagonals to the midpoint of the base of the square can be found to be the square root of five, using the Pythagorean theorem. Therefore, if the two diagonals to the midpoint of the base are extended along the base of the square, the length of the base of the new rectangle will be two times the square root of five, and the length of the sides will be two.

Therefore, the ratio of the length of the two longer sides to the length of the two shorter sides of this new rectangle will be $2\sqrt{5}:2$, or $\sqrt{5}:1$, or $\sqrt{5}$.

The geometry of the root five rectangle is the basis of the geometry of the golden rectangle. A golden rectangle can be found inside a root five rectangle in one of two ways. A root five rectangle can be formed from a square with two golden rectangles on the ends, or from two golden rectangles, which overlap each other to the extent of a square. The ratio of the length of the two shorter sides to the length of the two longer sides of a golden rectangle is .6180 . . . The ratio of the length of the two longer sides to the length of the two shorter sides of a root five rectangle is 2.236 . . . If the dimensions of two vertical golden rectangles are subtracted from the dimensions of a root five rectangle, the result is a square: 2.236 . . . - 1.236 . . . = 1.

Similarly, a golden rectangle can be constructed from a square, and a vertical golden rectangle, placed on its end. Therefore, a root five rectangle, also, consists of two golden rectangles, which overlap each other to the extent of a square.

> Out of the 120 Greek vases in the Boston Museum which can be subjected to a "dynamic" analysis, eighteen are based on the $\sqrt{2}$ theme (six having as overall frame the $\sqrt{2}$ rectangle, 1.4142, itself), six on the $\sqrt{3}$ theme (three having the rectangle $\sqrt{3}$ as frame); all the other ones on themes connected with Φ or $\sqrt{5}$. (Ghyka, 1977, p. 136)

The length of the diagonal of a square with sides of length one is the square root of two. Therefore, a root two rectangle can be constructed on the diagonal of a square. The length of the diagonal of a root two rectangle with ends of length one is the square root of three. Therefore, a root three rectangle can be constructed on the diagonal of a root two rectangle. The length of the diagonal of a root three rectangle with ends of length one is the square root of four. Therefore, a root four rectangle can be constructed on the diagonal of a root three rectangle. The length of the diagonal of a root four rectangle with ends of length one is the square root of five. Therefore, a root five rectangle can be constructed on the diagonal of a root four rectangle. Therefore, a spiral of root rectangles can be constructed with the diagonal of each root rectangle forming one of the two longer sides of each consecutive, and larger root rectangle. See Figure 101.

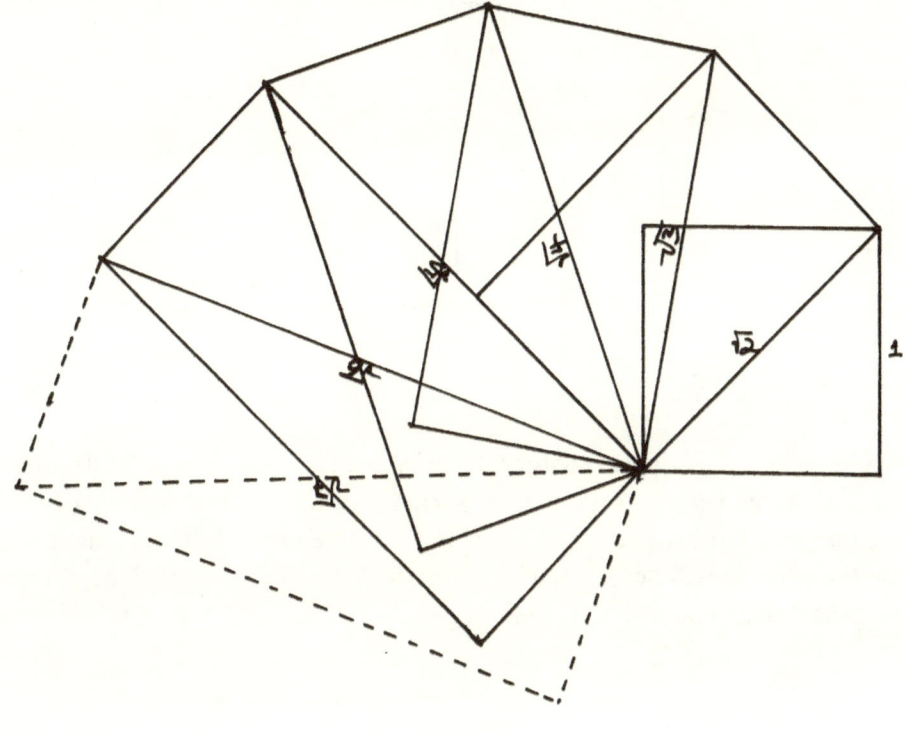

Figure 101

The root three rectangle was one of Plato's favorite figures, and was considered by him to be the most beautiful rectangle. The root four rectangle is, simply, a double square shape. It is one of the rectangles found, most commonly, in modern art, architecture, and materials' design.

A reciprocal rectangle of a rectangle is a similar, but smaller rectangle, constructed from one of the shorter sides, and equal segments of the two longer sides, inside the original rectangle. In order to find the ratio of the dimensions of the reciprocal rectangle, the ratio of the dimensions of the original rectangle is inverted. This is the reason for the name "reciprocal rectangle".

In order to obtain the reciprocal rectangle of a root two rectangle, one can divide the length of the two longer sides in two. Then, the length of the longer side of the reciprocal will be one, while the length of the shorter side will be $\sqrt{2}/2$. Therefore, the ratio of the dimensions of the reciprocal rectangle will be $1:\sqrt{2}/2$, or $\sqrt{2}$. See Figure 102.

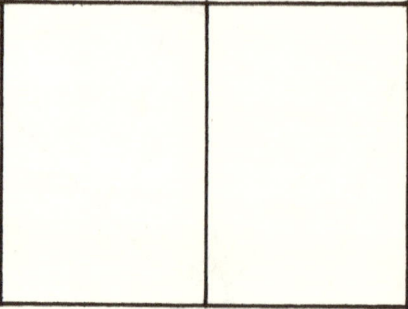

Figure 102

In order to obtain the reciprocal rectangle of a root three rectangle, one can divide the length of the two longer sides in three. Then, the length of the longer side of the reciprocal rectangle will be one, while the length of the shorter side will be $\sqrt{3}/3$. Therefore, the ratio of the dimensions of the reciprocal rectangle will be $1:\sqrt{3}/3$, or $\sqrt{3}$. See Figure 103.

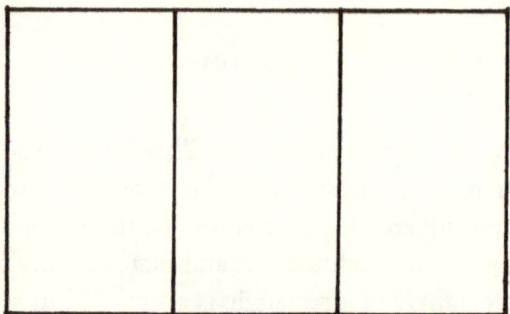

Figure 103

In order to obtain the reciprocal rectangle of a root four rectangle, the length of the two longer sides can be divided into four equal parts. Then, the length of the longer side to the reciprocal rectangle will be one, while the length of the shorter side will be $\sqrt{4}/4$. Therefore, the ratio of the dimensions of the reciprocal rectangle will be $1:\sqrt{4}/4$, or $\sqrt{4}$, or 2. See Figure 104.

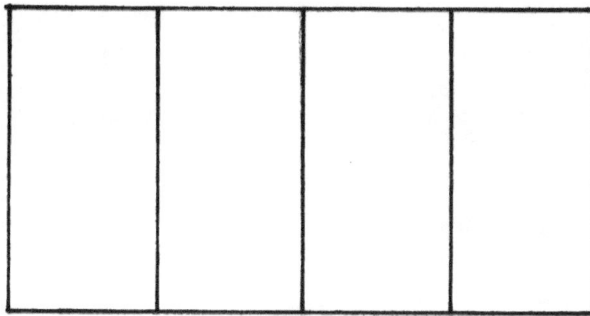

Figure 104

In order to obtain the reciprocal rectangle of a root five rectangle, the length of the two longer sides can be divided into five equal parts. Then, the length of the longer side of the reciprocal rectangle will be one, while the length of the shorter side will be $\sqrt{5}/5$. Therefore, the ratio of the dimensions of the reciprocal rectangle will be $1:\sqrt{5}/5$, or $\sqrt{5}$. See Figure 105.

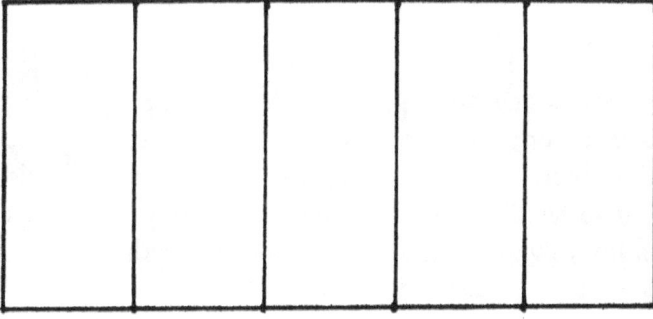

Figure 105

VI

The Golden Section in Art and Architecture

B. The Human Body

The book, *The Diagonal*, consists of twelve magazines, all entitled *The Diagonal*, published between November of 1919, and October of 1920. It was, mostly, written by Jay Hambidge, and published by the Yale University Press. Each magazine in this book contains articles on the analysis of the proportions of the bones of the human body, and the proportions of classic Greek vases, patera or bronze pans, and bronze mirrors, mostly, from the fifth and sixth centuries B.C. It, also, contains articles on phyllotaxis, or the golden section found in the architecture of plants, and art. It, also, contains articles on the different properties of the $\sqrt{2}$ and the $\sqrt{5}$ rectangles, both of which are found, commonly, as the overall design theme of these classic, Greek vases, patera or pans, and mirrors.

According to Hambidge, the rectangle, that the overall frame of the side elevation of average human skull fits into, consists of two root-five rectangles, placed together, vertically, and side by side. See Figure 106. Each of these root-five rectangles can be broken down further into what he calls five "whirling square" rectangles, or golden rectangles. This is the same rectangle, almost a square, that the grand span of the human body fits into. The grand span is the rectangle, formed by a human body, standing erect, and with arms, outstretched, horizontally. The height is measured from the top of the head to the base of the feet. The width is measured between the extremities of outstretched finger tips. It is equal to 2.236 divided by two, or 1.118.

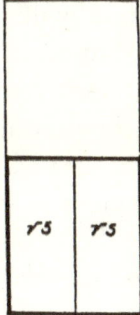 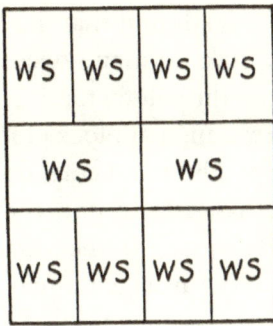

Figure 106

According to Hambidge, the rectangle, that the overall frame of the front elevation of the average, male, human skeleton of the torso and head fits into, is a vertical root-five rectangle. This root-five rectangle can be broken down further into a square, and two horizontal golden rectangles. One golden rectangle contains the rectangle, which contains the overall frame of the front elevation of the skull, and neck. The other golden rectangle contains the rectangle, which contains the overall frame of the front elevation of the skeleton of the pelvic bones. The square in the middle is the square, formed by the proportions of the bones from the top of the shoulder, or clavicle to the top of the pelvic bones, and the width of the shoulder bones. See Figure 107.

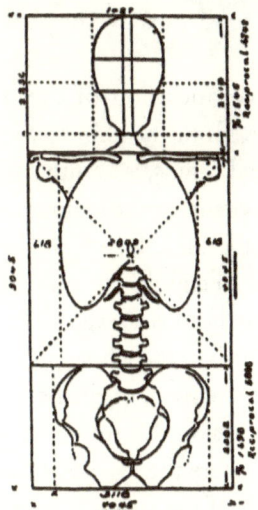

Figure 107

Hambidge, also, found that the rectangles, formed by the side elevation of the average, male, human skeleton contained a number of proportions, involving $\sqrt{5}$ and the golden ratio. He showed how these rectangles could be carved out of a rectangular block of marble to begin the sculpture of a typical Grecian statue of a human being, during the classical age of Greece.

Hambidge has, also, found proportions involving the square root of five, and the golden ratio in the rectangle containing the overall frame of the side elevation of the average human pelvic bone, and the relative lengths of the bones, making up the middle finger, and longest toe in a skeleton of an average human being.

In the book, *The Geometry of Art and Life* (1977, pp. 107 and 108), Matila Ghyka shows how Zeysing found that the ratio of total height to navel height in an average male adult is 1.60, or 8/5. The ratio of total height to navel height in an average female adult is 1.666, or 5/3. Both of these ratios are the ratio of the larger to the smaller of two consecutive Fibonacci numbers.

From the Canon of Polykleitus, Leonardo da Vinci determined that if a circle is circumscribed around the figure of a human being with hands and feet outstretched, the center of the circle would be the navel.

> Leonardo's translation: "The standards according to which all measurements are wont to be made, are likewise deduced from the members of the body; such as the digit, the palm, the foot and the cubit, all of which are subdivided by the perfect number which the Greeks called Teleios.
>
> "Nature in the composition of the human frame has so ordained that the face, from the chin to the highest point of the forehead where the head begins, is a tenth part of the whole stature; the same proportion obtains in the hand, measured from the wrist to the extremity of the middle finger; the head, from the chin to the top of the scalp, is an eighth. From the top of the chest to the highest point of the forehead is a seventh. From the nipples to the top of the scalp is a fourth of the whole stature. If the length of the face, from the chin to the roots of the hair, be divided into three equal parts, the first division determines the place of the nostrils; the second the point where the eyebrows meet. (The ear likewise is a third of the length of the face.) The foot is a seventh part of the height of the entire frame; the cubit and the chest, the width across the shoulders, are each a fourth.

"If a man should lie on his back with his arms and legs extended, the circumference of the circle which may be described about him with the umbilicus for a center, would touch the extremities of his hands and feet.

"The same obtains if we apply a square to the human figure; for, like the contiguous sides, the height from the feet to the top of the head is found to be the same as the distance from the extremity of one hand to the other, when the arms are extended." (Hambidge, 1920, p. 99)

Here, we have a close approximation of the circling of the square figure in the overall proportions of an adult human being.

"Vitruvius, architect, records in his work on architecture, that the measures of man are by nature distributed in this wise, that is: that four fingers make a palm, and four palms make a foot; six palms make a cubit; four cubits make a man; and four cubits make one step; and twenty-four palms make a man; and these measures constitute the human edifice." (Ibid.)

VI

The Golden Section in Art and Architecture

C. The Rock Tomb at Mira

Plate I shows the Egyptian temple plan of a rock tomb at Mira, as done by Ernst Moessel in *Die Proportion In Antike Und Mittelalter* (1926). Mira exists in Asia Minor, which includes "most of Asiatic Turkey: formerly called *Anatolia*." (*Webster's New World Dictionary of the American Language*, College Edition, s.v. "Asia Minor.") The design plan is similar to the design plans of many other ancient, Greek, four-sided buildings.

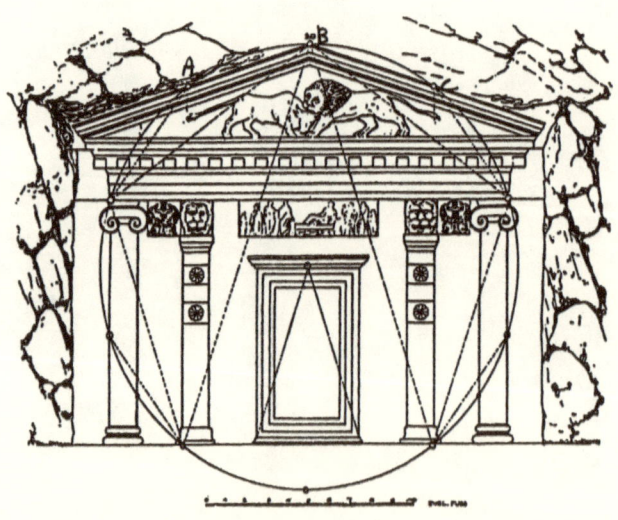

Plate I

Basically, its structure consists of a circle, which circumscribes a regular decagon, whose radii form the various subdivisions of the design of the facade. The width of the facade is equal in length to the length of the diameter of the circumscribing decagon. The ratio of the width of the facade at the base of the timberwork construction to the height of the facade is equal to the ratio of the length of the diagonal of a regular pentagon to the length of its respective height. The whole height of the facade is divided into golden section at the base of the timberwork construction. The top part of the timberwork construction is divided into a pattern of nineteen alternate squares. The four rosettes on the two inside pillars of the facade each contain ten petals.

The isosceles triangle formed by the apex of the roof, and the two vertices of the regular decagon, that intersect at the base of the facade, is a golden triangle. The isosceles triangle formed by the midpoint of the top side of the rectangle formed by the outside frame of the doorway, and the two base vertices of this rectangle is a golden triangle. This midpoint, also, is the center of the decagon, and circle, that circumscribe the facade.

Each of the two top halves of the roof structure intersect the circumscribing circle at a point, which is one vertex of the circumscribing decagon. The angle, cut out by the apex of the roof, therefore, is 2/5 of a circle, or 144°. This can be proven, as follows. Consider the diagram in Plate I.

$$\angle ABC = (1/2)(8/10)360°$$
$$= (4/10)360°$$
$$= (2/5)360°$$
$$= 2*72° = 144°$$

Therefore, the slope of each side of the roof is 18°.

Again, this is the squaring of the circle motif, because the square, which circumscribes, or forms the frame for the building, also, circumscribes the circle, which forms the various partitions of the building.

VI

The Golden Section in Art and Architecture

D. The Parthenon

The Parthenon was built between 447 B.C. and 438 B.C. It was dedicated to the Greek goddess, Athena Parthenos, "the Virgin" (Stillwell, "Parthenon", 1966, p. 412). It is located on the Acropolis in Athens, Greece. It was designed by the two Greek architects, Ictinus and Callicrates, under the supervision of the sculptor Phidias.

According to Hambidge, the dimensions of the ground plan, as fixed by the lowest step, are 238.003 by 111.31 British feet. These dimensions reduce to the ratio 2.1381996. The rectangle, so formed, is composed of two squares, and two root-five rectangles. See Figure 108.

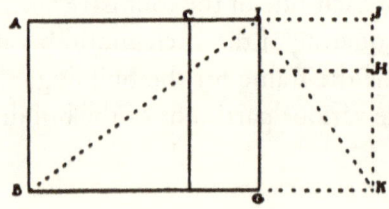

Figure 108

The rectangle IK is the reciprocal of the rectangle AG.

The two shorter sides have eight columns. The two longer sides have seventeen columns. The four angle, or corner columns are thicker in diameter than the others. The columns on the two shorter sides are spaced at a, slightly, greater distance apart than the columns on the two longer sides. All of the columns taper, slightly, towards their tops.

Figure 109 shows an illustration from the book, *The Parthenon* (1924), by Jay Hambidge. It is a summary of his analysis of the proportions of the facade of the Parthenon, according to dynamic symmetry. The complete analysis of the facade with diagrams is given on pages 14 thru 19 of this book (Chapter Two).

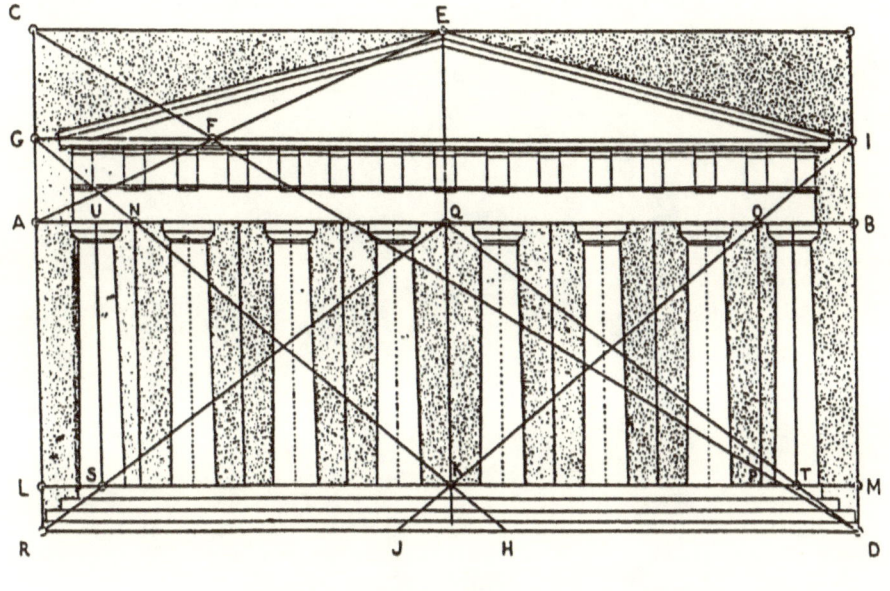

Figure 109

The ratio of the overall rectangle is 4/2.3416408, or 1.7082039. This is equal to $3\sqrt{5}$ - 5. Areas GH and IJ are each composed of two golden rectangles. Area GM is composed of four golden rectangles. Area NP is similar to area GM, and is, also, composed of four golden rectangles. This area, exactly, contains the six rectangles containing the six internal columns.

Area AD is composed of four 1.4472 rectangles, standing vertically. This ratio is derived from the combination of a square with a root-five rectangle placed, horizontally, on top $(1+1/\sqrt{5} = 1+.4472 = 1.4472)$. Area UT is a similar rectangle to area AD. Areas CQ and EB are both horizontal root-five rectangles.

The total height of the facade, CR, is divided into golden section at point A by line segment AB. Therefore, the ratio of the area AD to the area CB is equal to the golden ratio. This, also, is the same for the areas, divided by the top of the pillar lines, for the two flank elevations.

The rectangle, formed by the two flank elevations, is composed of a double, or two facade rectangles, plus a vertical .236 area. The .236 area is equal to a root-five rectangle minus two squares (2.236-2 = .236). It is found, elsewhere, in the overall design theme. A more detailed examination of the base rectangle, and flank elevations is given in Chapter One of Hambidge's book, *The Parthenon*.

Using a protractor, I have found, that the angle of the roof, as shown in this diagram, is 15°. 360° divided by 15° is 24. 360° divided by 2*15°, or 30° is 12!

VI

The Golden Section in Art and Architecture

E. The Cathedral of Notre Dame

The Cathedral of Notre Dame exists in Paris, France. It was built between the years of 1163, and 1263 A.D. It was one of the first of many cathedrals built, during the Gothic Age. The Gothic Age lasted "from the middle of the 12th century to the early 16th century" (*Webster's Seventh New Collegiate Dictionary*, 1963, s.v. "Gothic"). It served as a model for other cathedrals built in western Europe, including, in France, Germany, England, Italy, and Spain. Notre Dame are French words, and mean "our lady".

Figure 110 shows the design analysis of the facade of the Cathedral of Notre Dame. The overall frame of the circumscribing rectangle is a golden rectangle. It is divided by a square, close to the base, which circumscribes a circle, which circumscribes a regular decagon, whose partitions form the design analysis of the facade.

Figure 110

All of the Gothic mason's marks are a geometric figure, drawn on a square, triangular, or circular grid, inscribed inside a circle. They are all based on either the semicircle and its diagonal, or the squaring of the circle with inscribed triangle, or the circling of the square figures.

> We shall not be surprised to see in the next chapter that, according to the converging results of recent researches about Canons of Proportions and plans used by Gothic Master Builders, the fundamental Gothic Diagram, the Key-Diagram transmitted from Master to Master (the third degree of initiation in the craft) was based on pentagram and decagon, placed within the circle of orientation of the church or cathedral. (Ghyka, 1977, P. 119)

This "Key Diagram" was, undoubtedly, used in the conception of the design of the facade of the Cathedral of Notre Dame. The image of the pentagram, inscribed in the central circle, which radiates from the rose window, which is inscribed in the central square, which is inscribed in the central golden rectangle, can, easily, be seen from the analysis, given us by

Professor Moessel, of the facade of the Cathedral of Notre Dame. This "Gothic Master Diagram" (Ibid., p. 121) is shown in Plate II.

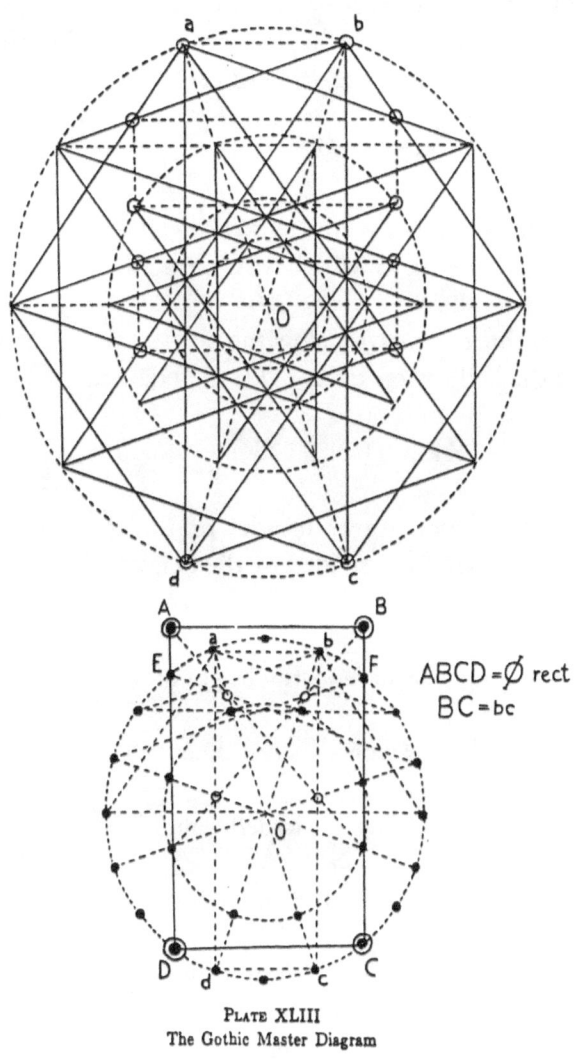

ABCD = $\emptyset$ rect
BC = bc

PLATE XLIII
The Gothic Master Diagram

Plate II.

The overall fame of the facade, as outlined by the dotted lines in Figure 110, is a golden rectangle. The total height of the facade is 66.66 meters. The total width of the bell tower facade is 41.20 meters. The height divided by the width produces the ratio 1.618. Further subdivisions of this golden rectangle can be found in the overall frames of the doors, and archway designs. The external golden rectangle can be broken down into nine equal, and symmetrical golden rectangles.

The distance, or gap between the bell towers is equal to the width of the facade of the bell towers times the value of (1/2)*0.600. That is,

$$41.20*(1/2)*0.600 = 12.36$$

Also, the distance, or gap between the bell towers is equal to the length of one side of the regular decagon, which is circumscribed inside the circle, that surrounds the rose window.

The height of the lower part, below the bell towers, is equal to the width of the facade of the bell towers times $(1/2)\sqrt{5}$. That is,

$$41.20*(1/2)*2.236 \ldots = 46.06$$

The width of the bell tower facade, or the diameter of the circle, surrounding the rose window, divided by the height of the bell tower facade is 2. That is,

total height - height of the lower part = height of the bell tower facade, or

$$66.66 - 46.06 = 20.60$$

Therefore, width of the bell tower facade/height of the bell tower facade =

$$41.20/20.60 = 2$$

The triangle, formed by the width of a line segment, drawn one meter below, and parallel to the base, and the two line segments, drawn from the endpoints of this line segment to the center of the rose window is a 72°, 54°, 54° triangle. This triangle is similar to a triangle formed by two consecutive radii, and one side of a regular pentagon.

The triangle formed by the two line segments, going from the central dot at the top of Figure 110 to the base of the bell tower facade, and the line segment formed by a central segment of the base of the bell tower facade is a golden triangle. It is part of, and is similar to the triangle, formed by the two line segments, going from the central dot at the top of Figure 110 to the two endpoints of the line segment, drawn one meter below, and parallel to the base, and this same line segment. This triangle is, also, a golden triangle.

All of the open arches above the doors, in the windows on the second story, and in the bell towers are based on the shape of two semicircles, which overlap each other to the extent of an equilateral triangle. That is, an equilateral triangle can be inscribed inside these arches, such that it touches the two base vertices, and the apex of the respective arch. Then, the two arcs of the two lateral sides of the equilateral triangle can be formed from equal segments of circles, which are drawn from centers at the respective, opposite base vertex of the respective arch. This shape is, also, found in dome construction.

The structures on the sides of the facade are buttresses, or abutments, which are used to support the tall building, especially, as the arches put a sidewise pressure, or thrust on the general building structure. Flying buttresses are placed at the back, or along the two lateral sides of the main building to support this pressure. The general shape, or floor plan of all cathedrals, built during the Gothic period, is that of a crucifix. The horizontal cross section is similar to the horizontal cross section of all cathedrals, built during this period.

Within the rose window of the facade of the Cathedral of Notre Dame, there are twenty-four partitions, twelve on the top half, and twelve on the bottom half. There are twelve cycles of the moon in one solar year. There are twenty-four hours in a day. This, perhaps, indicates, as with the Court of the Lions, the, purposeful, implementation of this astronomical knowledge by the French masons into the architecture of the facade of the Cathedral of Notre Dame.

This architectural design displays the importance of the concept of the squaring of the circle. The center of the rose window is at the center of the design with all points radiating out from it, and being extensions of it. The overall frame of the design is formed by the square, which is found to circumscribe the circle, which radiates to the edges of, and pinnacle of the building.

VI

The Golden Section in Art and Architecture

F. The Palace of the Alhambra

The Palace of the Alhambra is an ancient Moorish castle. It was built between 1238 and 1358 A.D., and by the last Moorish rulers of Spain. It rests high atop the hills, surrounding Granada, Spain in southern Spain. The word Alhambra is a Spanish word, and comes from the Arabic words "al-hamra'" (*Webster's Seventh New Collegiate Dictionary*, 1963, s.v. "Alhambra"), which mean "the red house" (Ibid.). The reason for this name is derived from the fact that the outside walls of the palace are covered with red tile.

The Moors ruled in Spain from the eighth century to the thirteenth century A.D. They had a strong cultural influence on the Spanish population, including with food, language, the arts, and architecture. As the Christian forces of the north grew stronger, the Moors were pushed to the south. This is why the Palace of the Alhambra was built in southern Spain. It was the last stronghold of Moorish rule in Spain. It was not until 1492 when Queen Isabella of Castile, and King Ferdinand II of Aragon invaded and conquered Granada, that Moorish rule came to an end. The last Moorish king was Boabdil.

The Palace of the Alhambra consists of ten sections: the Abencerrajes Gallery, the Apartments of Charles V, the Barca Gallery, the Court of the Lions, the Court of the Myrtles, the Garden of Daraxa, the Garden of Machuca, the Hall of the Ambassadors, the Palace of Charles V, and the Royal Baths. The Apartments and Palace of Charles V were built after, or added onto the original Moorish structure.

In the book, *Die Proportion in Antike und Mittelalter*, by Ernst Moessel (1926), the proportions of the Court of the Myrtles are given: 37.20 m.

x 22.95 m. The length of this court divided by its width yields the ratio 1.620915, a very close approximation of the golden ratio. It is Moessel's theory that the proportions of this court were designed to fit the proportions of a golden rectangle. Also, it has been found that the proportions of the rectangles, that form the overall frames of many of the arches in the palace, form the proportions of a golden rectangle. The Court of the Myrtles is located at the center of the entire structure of the Palace of the Alhambra. Myrtles is the name of a flower, that grows, commonly, in southern Europe.

In the Court of the Lions, there is a fountain with statues of twelve lions at the base. They face outwards, standing poised with water running out of their mouths. It is possible that the number twelve in this case represents the twelve cycles of the moon within one solar cycle. The Moorish people were a group of nomads, that roamed northern Africa, that had Arabic stock. This fact of astronomical lore was, obviously, still important to them.

APPENDICES I.-IV.

Appendix I

The Sumerians: The Ancient Time Tellers

The Sumerians established the first civilization of man, which is, sometimes, known as the "cradle of civilization" (ca. 4,000-3,000 B.C.). They lived at the mouth of the Tigris and Euphrates rivers in what is, now, modern day Iraq. They were the inventors of a written language, called Cuneiform, the sundial, a base six number system, our modern system of time telling, and a 360 degree circle. Their system of time telling included a 60 minute hour, a 60 second minute, and a 24 hour day.

They used simple geometry and multiplication tables. The Sumerians advanced the sciences of astronomy and medicine, codified their laws, which were written on clay tablets, used handwoven fabrics and metals, including copper and tin, fused into bronze, and gold and silver objects, and built elaborate palaces and temples. They were the first civilization of man to plant cultivated crops, irrigate, and pull plows with beasts of burden to plow, and harvest these crops.

Being so close to the equator, there was little difference in the temperature between seasons. Because of this, the Sumerians had little need for a solar calendar, and used a lunar one, instead. It is, perhaps, for this reason, that they placed such great emphasis on the number twelve. For example, they used a base six number system. That is, there are six lunar cycles between consecutive solstices in the year.

> Next to the sun, the most noticeable object in the sky is the moon, and it undergoes a periodic change, almost as noticeable as day and night, in the form of its changing phases. From a thin crescent immediately after sunset (the "new moon"), it enlarges

and moves farther from the sun, till it is a full moon rising at sunset; then it shrinks and moves closer to the sun until it is a thin crescent barely visible just before sunrise. Soon after this, another new moon appears in the sky just after sunset. These phases arise as a result of the moon's revolution about Earth and its changing position, therefore, relative to the sun.

The entire cycle is completed in 29.53 days, and to those who first worked out the period, this meant that the interval between new moons was sometimes twenty-nine days and sometimes thirty days. This interval was the "month," and lunar calendars have been based on this month for thousands of years. The religious calendar of the Jews and Moslems, even today, is lunar in nature.(Asimov, 1982, p. 184)

It is easier to mark off the cycle of the seasons by counting the months than by counting the days. While the seasons are not mathematically regular, as are the day, the week, and the month, they are a most essential cycle, whether human beings are food gatherers, hunters, or farmers. It was eventually the experience of human beings that the season cycle, or "year," which marked the period of the revolution of Earth about the sun, was a little over twelve lunar months long.

Actually, the year is 12.37 lunar months long, and the Babylonians worked out a pattern of years, sometimes twelve months long and sometimes thirteen, a pattern that repeated itself every nineteen years. This was adopted by the Greeks and the Jews (and is still used today in the Jewish religious calendar).

The Egyptians had a simpler seasonal cycle and depended almost entirely on the one annual event of the flooding of the Nile, which came, on the average, every 365 days. They set 365 days to the year, therefore, and filled it with twelve months of thirty days each, followed by five days of celebration. The months did not fit the phases of the moon, but the Egyptians didn't mind that. This was the first solar calendar.

The Romans eventually adopted the Egyptian calendar in 44 B.C. and, with the help of the Greek astronomer Sosigenes, added the additional refinement of a leap day. Since the year was actually 365 1/4 days long, every fourth year was given 366 days.

As a matter of fact, the year is not exactly 365 1/4 days either, but is 365.2422 days long—a trifle shorter. This means that three times every four centuries, a year that would ordinarily be a leap year should not be. The necessity for this was first pointed out by the English scholar Roger Bacon (1220-92), but it is very difficult to change a calendar at any time. The change was not carried through successfully until Pope Gregory XIII (1501-85) decreed it in 1582, and even then, only Catholic Europe followed him at first. Nevertheless, the new "Gregorian calendar" spread, and it is now worldwide. (Ibid., pp. 184-185)

Sixty is equal to six times ten, or twelve times five. The first division combines the base six and the base ten number systems. Perhaps, the Sumerians noticed the ease of counting on ten fingers, or ten toes. Twelve is the number of lunar cycles in a year. Five is the number of the pentagram.

Twelve times 29.53 days is 354.36, or, approximately, 354 days. This is the number of days in twelve lunar cycles. The number of days in one solar cycle is 365.2422, or, approximately, 365.25 days. The average between 365.2422 and 354.36 is 359.8011, or almost, exactly, 360 days! This is, probably, why the Sumerians chose the number 360, as the number of degrees in a circle. The circle could have represented the circumference of the earth; 360 degrees, the number of days on average between the time it takes the earth to revolve around the sun, and for the moon to revolve around the earth twelve times.

The Sumerians, and other ancient civilizations could measure the length of a solar year, closely, by counting the number of days between consecutive solstices, or longest days and shortest days with a sundial, or some other type of solar measuring device. The sun would be, directly, overhead, two days every year. 360 is equal to six times sixty.

In the British foot, which is known to have been derived from ancient times (perhaps, the Egyptians), there are twelve inches with four divisions, each. Twelve could represent the number of approximate lunar cycles in a year. The four divisions could represent the four phases of the moon within each cycle. Also, there are four seasons in a year. There are four solar years in every leap year cycle. There are 16 fluid ounces in a pint, two cups in a pint, 12 or 16 ounces in a pound, two pints in a quart, and four quarts in a gallon. Also, there are four quadrants, or 90° (right) angles in a circle. Time is considered by many to be the fourth dimension.

Appendix II

The Egyptians Brought Culture to Central America and Peru

It is known that the Egyptians traveled in papyrus reed, or some other type of boats to the New World. In Thor Heyerdahl's Ra I and Ra II voyages, he proved that it was possible to travel in a barge or an ark, made of papyrus reeds, bundled together, from Safi on the northwest coast of Morocco, across the Atlantic Ocean, and to Barbados, where he landed, or the northeast coast of South America.

On May 25th, 1969, Thor Heyerdahl and a crew of seven set forth on their Ra I voyage. The boat was poorly built, however, and after 55 days, and still 600 miles from Barbados, the boat began to fall apart, and the expedition had to be abandoned. On May 17th, 1970, Thor Heyerdahl and a crew of eight set forth on their Ra II voyage. This time, the boat, being more sturdily constructed, landed at Barbados, 3,270 miles and 57 days, later.

> To summarize, all available evidence shows that it was the papyrus ship which had developed all the seagoing ship's characteristic properties and which subsequently became the model for the wooden ship, not the other way around. The design of the papyrus ship was already developed when the First Dynasty began building pyramids along the Nile. (Heyerdahl, 1978, p. 12)

The Egyptians could have used a sundial, cemented to the front, middle or some other part of the boat, where the sun shines, constantly, during the day, to determine the exact latitude and time of day at which they were sailing. Either that, or the Egyptian sailors could have used the shadow thrown by the mast of the boat to determine the exact latitude and time of day. Although, of course, the shadow thrown by a mast would, probably, extend way beyond the edge of a boat. By, constantly, orienting themselves to the same latitude each day, and each hour of the day, the Egyptians could sail in a true line from east to west.

They would know, that if they would not encounter a land bridge on their trip, they would, surely, travel from one side of the world to the other, perhaps, landing back in Egypt, or at least in the ancient East. Probably, during the night, the ancient seafarers used the stars to guide their way if not to let their mast down, and drift, while they slept.

> Numerous theories of voyagers drifting from Africa to tropical America have been proposed to explain the sudden blossoming of high culture from Mexico to Peru. Like the ancient peoples of the Old World, Indians of the Americas worshiped the sun, built pyramids and giant stone statues, married brother to sister in royal families, wrote in hieroglyphs, performed cranial surgery, and mummified the dead. (Heyerdahl, Jan., 1971, p. 46)

Papyrus is still grown around Lake Titicaca, high in the Andes Mountains in Peru. It is used to construct simple huts for living, and small boats.

> The evidence is abundant. In a tomb relief, ancient Egyptians carry papyrus to builders of a reed boat. In verses from the Bible, in scenes found at Nineveh, in writings of the Roman historian Pliny, the reed boat stands as one of man's most ancient vessels. And, strangely, it survived into recent times in Mesopotamia, Ethiopia, Sardinia, Corfu, and Morocco. Fishermen still ply central Africa's Lake Chad in reed boats. (Ibid.)

There are other elements, which lead to the belief that the American Indians acquired knowledge, and, almost, the entire foundation of their culture from the Egyptians. The American Indians, especially, the Aztecs, like the Egyptians, believed that the sun was God, or at least, should be worshiped as a god.

> The Aztecs believed that the gods in their turn created the earth. The most important act in this creation was the birth of the sun. The sun is supposed to have been born at Teotihuacan through the

self-sacrifice of a little leprous god. The remaining gods followed
his example of sacrifice to provide the blood needed to set the sun
moving across the sky.

In order to keep the sun moving on its course it had to be
fed every day with human blood. The Aztecs regarded sacrifice as
a sacred duty towards the sun. Without this the life of the world
would stop. Therefore constant human sacrifices—mainly of war
captives—had to be provided. It is thought that more than 20,000
were slain each year. (Bankes, 1982, p. 54)

Also, the Aztecs had a sun calendar, and an almanac, called tonalpohualli,
of 260 days. The Aztecs used a year of 365 days, composed of eighteen
months with twenty days each, and "the final 5 unlucky days" (Thompson,
1966, vol. 4, p. 628). This Indian calendar bears a strong resemblance to the
Egyptian solar calendar, which had 360 days, composed of twelve months with
thirty days each, and a final five days, which were reserved for celebration.
According to the *Webster's New World Dictionary of the American Language*
(College Edition, 1966, s.v. "almanac"), an almanac is "a yearly calendar of
days, weeks, and months, with astronomical data, weather forecasts, tables
of useful information, etc." Notice, the 260 days in the almanac is, exactly,
100 days short of the 360 day year, or five months of twenty days.

Like the Egyptians, the early American Indian tribes established nature
worship, as the center of their religion. Also, some of the more important
Indian tribes, such as the Mayans, had established a system of hieroglyphics,
similar to that of the Egyptians.

There is one pyramid, which has been found on the central Yucatan
peninsula, built by the Mayans about 100 A.D., called "Cinco Pisos". In
Spanish, this means five stories. It has a stone column on a flat surface in front
of the pyramid, which predicts, exactly, when the sun is, directly, overhead.
This pyramid is located just below the Tropic of Cancer, where the sun shines,
directly, overhead twice a year.

The column is tapered at the top, and a flat, circular stone sits on top of
it. Only twice a year, when the sun is, directly, overhead, does the circular
stone, sitting on its top, cast a shadow, which, directly, covers the entire surface
of the column. Surely, this astronomical device, or sun gnomon could have
been used to determine the exact length of a solar year, perhaps to the nearest
hour and minute. Also, it has been found that this pyramid was built with
other astronomical abilities.

Lastly, all of the pyramids, built by the Aztec and Mayan Indians, were step pyramids. The earliest Egyptian pyramids were step pyramids. For example, the first pyramid, built in Egypt, that of Zoser at Saqqarah, was a step pyramid. Zoser Neterikhet was the second king of the third dynasty (c. 2686-c. 2613 B.C.). Obviously, the simpler, Central American pyramids, which were built, need not have planed sides. Instead of having pointed tops, they were built with temples at the top, which were designed as places to sacrifice animals and human beings.

The earliest Indian civilization of any significance, known to archeologists, in Central America is the Olmec civilization. This civilization flourished from about 1200 to 100 B.C. 1200 B.C. was about the same time as the downfall, or end of the golden civilization in Egypt. The largest ceremonial site, built by the Olmecs was at La Venta. La Venta is located southeast of Veracruz in southeastern Mexico, and on the Gulf of Mexico. They built a 110 foot tall clay pyramid there.

> Between about 1200 and 100 BC a people, called the Olmecs by archaeologists, lived on the Gulf Coast of Mexico. They built ceremonial centres such as La Venta which were carefully planned temple communities. At La Venta a group of buildings was arranged symmetrically along one axis. At the south end was a clay pyramid, shaped like a fluted cone.
>
> Olmec religion seems to have centered around the jaguar in various guises. At La Venta three identical mosaic pavements, made from blocks of green serpentine, were discovered. Each was laid out in the form of a stylized jaguar face, and as soon as it was finished it was covered up. These could have been some kind of offering to the beast. (Bankes, 1982, pp. 50 and 52)

However, modern field investigations have shown that an earlier Indian civilization existed one thousand years before the Olmecs. It existed in the Central Andes of Peru. "The first large architectural complexes were raised in the Central Andes between 2500 and 1500 B.C., drawing upon thousands of years of local cultural developments." (Quilter, 1985, p. 931) "The sites under discussion were huge constructions made of tons of stone, adobe, and fill formed into platform mounds, circular sunken plazas, terraces, and other public works." (Ibid., p. 932) This time period is known as the Preceramic Period.

When and who these early culture carriers were is not, exactly, known. However, due to the similarity of cultures between the Middle American Indians and the Middle East, it is assumed that the culture was brought from the Middle East. The Middle Eastern civilizations, especially Egyptian, had flourished, slightly, before, or at the same time.

Thor Heyerdahl has proved that the Peruvian Indians sailed to many of the South Pacific islands, and, possibly, to India, establishing, partly, Indian populations in these locations. Thor Heyerdahl had discovered in reading history books on the Peruvian Indians, that the ancient Peruvians had built large, balsa log rafts with sails, that could be navigated in the oceans. He had known that the currents off the western coast of South America first flow northward, and then westward. He had, also, known that the populations of many of the South Pacific islands had legends, and genetic inheritance, telling of ancestry from South America, including Easter island.

So, in order to prove his theory, he built a reconstructed model of one of these Indian rafts, called Kon-Tiki. He sailed from Lima, Peru to Tahiti. He thus proved that it would be possible for the Peruvian Indians to make such long and arduous journeys in pre-Columbian days.

In digging in the ruins of the ancient Toltec capital of Tollan, near Tula, Mexico, Charnay found several elaborately carved masks of Caucasian, Greek, Chinese, Japanese, and Negro faces. Desire Charnay was an Americanized French school teacher in Louisiana. He first decided to explore the ruins of ancient Mexico, during his summer vacations. Later, he was sponsored by the Lorillard Company. Also, many pyramids, and other monumental structures in Central America have oriental, Indian, Greek, and Assyrian styles and motifs. This evidence indicates that not only Egyptians, but, also, travelers of many other races and nations visited Central America.

Appendix III

The Logarithmic Spiral

There are many different types of spirals. These include the equiangular spiral; $r=a\theta^n$, including the spiral of Archimedes; the sinusoidal spirals; Euler's spiral; and Cotes' spirals.

The two most, commonly, known spirals are the spiral of Archimedes, and the equiangular spiral. The spiral of Archimedes is well known, because it is simple to construct, and is, virtually, mathematically, perfect. Its equation is found from the polar coordinates r and θ. Its formula is $r=a\theta$, where r is the length of the radius, and θ is the given angle at which the radius is rotated around the origin. a is a constant.

The spiral of Archimedes was discovered by Conan. Archimedes lived from 287 to 212 B.C. Archimedes was a Greek mathematician, philosopher, and engineer. He lived on the island of Sicily, off the southern most coasts of the boot of Italy. He was stabbed in the side with a spear when the Romans invaded a public bath, he was attending.

Archimedes was a lonely sort of eagle. As a young man he had studied for a short time at Alexandria, Egypt, where he made two life-long friends, Conon, a gifted mathematician for whom Archimedes had a high regard both personal and intellectual, and Eratosthenes, also a good mathematician but quite a fop. These two, particularly Conon, seem to have been the only men of his contemporaries with whom Archimedes felt he could share his thoughts and be assured of understanding. Some of his finest work was communicated by letters to Conon. Later, when Conon died,

> Archimedes corresponded with Dositheus, a pupil of Conon. (Bell, 1965, p. 30)

> Three of Archimedes' extant works are devoted to plane geometry. They are *Measurement of a Circle*, *Quadrature of the Parabola*, and *On Spirals* . . . . The third work contains 28 propositions devoted to properties of the curve today known as the spiral of Archimedes and which has r=Kθ for a polar equation. (Eves, 1976, pp. 136-7)

The logarithmic spiral was first discovered by René Descartes. He discovered it on the basis of "a study of dynamics" (Yates, 1974, p. 206). René Descartes was a famous French philosopher, mathematician, and physicist. He was born in La Haye, France in 1596. As a young man, he fought in many European wars. He invented the science of analytic geometry, or applying algebra to geometry, and vice versa. He reinvented the X, Y axes after the Greeks, and Cartesian coordinates. He died in Stockholm, Sweden in 1650.

The "properties of self-reproduction" (Ibid.) were worked out by Jacob Bernoulli (1654-1705). "It was Jakob Bernoulli in the 17th century who began to suspect that the golden mean was intimately connected with forms and patterns in nature." (Hoffer, 1977, p. 107)

> Bernoulli named it the logarithmic spiral. He noted that any line drawn from the center of the spiral will intersect it at precisely the same angle as any other line drawn from the center. For this reason it is also known as the equiangular spiral. Bernoulli was so impressed with it that he ordered it engraved on his tombstone. (Ibid., p. 108)

"Any [plane] curve proceeding from a fixed point (which is called the pole), and such that the arc intercepted between any two radii at a given angle to one another is always similar to itself, is called an equiangular, or logarithmic, spiral." (Thompson, 1942, p. 757)

The logarithmic spiral is unique, because it is found, commonly, in the growth of living things, and in physics. For example, it is found in the curve of an ocean breaker, as it is about to hit the shore. It is found in the curve of the tail of a comet, as it blows away from the sun, as the comet hurls around the sun.

In a slightly different, but closely cognate way, the same is true of the spirally arranged florets of the sunflower. For here again we are regarding serially arranged portions of a composite structure, which portions, similar to one another in form, *differ in age*; and differ also in magnitude in the strict ratio of their age. Somehow or other, in the equiangular spiral the *time-element* always enters in; and to this important fact, full of curious biological as well as mathematical significance, we shall afterwards return. (Ibid., p. 752)

The epeira spider, always, spins its web in the form of a logarithmic spiral. "Some webs, more than others, exhibit aspects of the Fibonacci sequence in their spiral and radial threads." (Hoffer, 1977, p. 106)

Shellfish from among the earliest forms of life exhibit a wide variety of logarithmic spirals. "It is possible to trace, even back in the plankton phase, distinct expressions of spiral organization associated with Fibonacci ratios," wrote A. H. Church. The globigerinae, planorbis vortex, terebra, turritellae and trochids all build their tiny bodies according to the logarithmic spiral, as does the snail.

One of the most spectacular sea spirals is created by the chambered nautilus. As the animal grows, it builds ever larger compartments in an expanding spiral formation. When each new apartment is ready for occupancy, the animal crawls forward, shutting off the previous compartment behind it with a layer of mother-of-pearl. The old living quarters remain filled with gas and air, so that the whole structure remains buoyant in spite of its massive build. Not only does the nautilus build the outside of its shell in a spectacular spiral, the inner partitions are also curved in the same graceful form. (Ibid., pp. 108-109)

"Why do bacteria grow at an increasing rate that may be plotted along a logarithmic spiral?" (Ibid., p. 111)

Other examples of how logarithmic spirals are found in nature are cited in the beginning of Chapter IV., Section B., "The Golden Spiral of the Golden Rectangle".

Appendix IV

The Five Regular Solids

A regular solid is a three-dimensional, closed figure with a finite number of faces, formed from regular and equivalent polygons. There are five regular solids, which exist, and which have been known since Plato's time: the regular tetrahedron, the regular octahedron, the cube, the regular dodecahedron, and the regular icosahedron. Plato was born in Athens, Greece around 427 B.C., and died there around 347 B.C.

> Fifthly, it was claimed that Pythagoras discovered the construction of the five regular solids. It was more probably Theaetetus who (as we read elsewhere) discovered the octahedron and the icosahedron; but the Pythagoreans were clearly acquainted with the pyramid or tetrahedron and the dodecahedron. The construction of the dodecahedron requires that of a regular pentagon, which again depends (as in Eucl. iv, 10, 11) on the problem of Eucl. ii, 11, about the division of a line in extreme and mean ratio, a particular case of the application of areas. The assumption that the Pythagoreans could construct a regular pentagon is confirmed by the fact that the pentagram, the triple interwoven triangle, or the star pentagon, was used as a symbol of recognition between the members of the school and was called by them health. (Heath and Lloyd, 1966, vol. 18, p. 906)

Theaetetus was born in Athens, Greece around 417 B.C., and died there in 369 B.C. "He studied the five regular solids of Plato and may have been the first to demonstrate that there were, in fact, only those five and that no other regular polygons could exist." (Asimov, 1982, p. 17)

The regular tetrahedron is a regular solid, which has four faces, which consist of equilateral triangles. See Figure 111. The regular octahedron is a regular solid, which has eight faces, which, also, consist of equilateral triangles. See Figure 112. The cube is a regular solid, which has six faces, which consist of squares. See Figure 113. The regular dodecahedron is a regular solid, which has twelve faces, which consist of regular pentagons. See Figure 114. The regular icosahedron is a regular solid, which has twenty faces, which consist of equilateral triangles. See Figure 115.

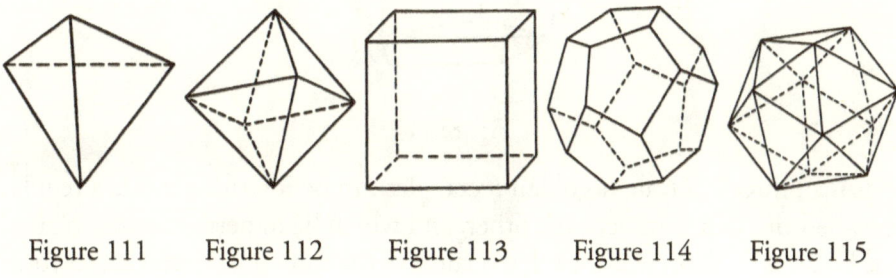

Figure 111   Figure 112   Figure 113   Figure 114   Figure 115

Two pairs of interlocking, regular solids can be constructed, such that their edges, and vertices intersect in such a way, as to make lengths, that are in golden section. A regular icosahedron can be inscribed inside a regular octahedron, such that each of its vertices divides one edge of the regular octahedron into golden section. See Figure 116. Notice, an octahedron has twelve edges, while a regular icosahedron has twelve vertices.

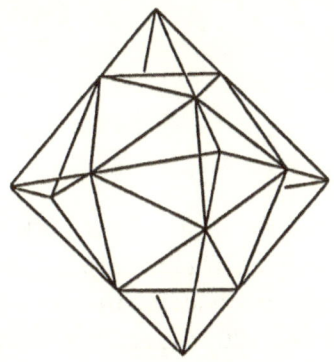

Fig. 116

A cube can be inscribed inside a regular dodecahedron, such that each of its edges forms one of the diagonals of one of the faces of the regular dodecahedron. See Figure 117. Notice, a cube has twelve edges, while a regular dodecahedron has twelve faces.

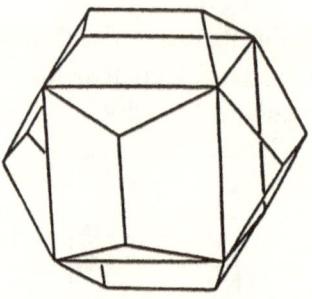

Figure 117

Also, three equivalent golden rectangles can be inscribed inside a regular icosahedron, which bisect each other, and which lie in perpendicular planes. See Figure 118. Notice, the twelve vertices of the three golden rectangles form the twelve vertices of the regular icosahedron.

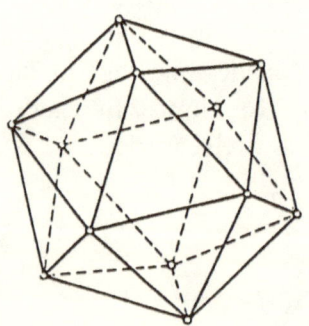 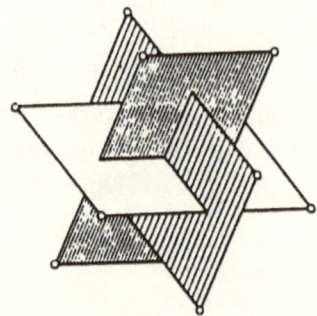

Figure 118

A Petrie polygon is a two-dimensional perspective of a three-dimensional figure with opposite vertices lying in a line perpendicular to the plane of vision. Petrie polygons of the regular dodecahedron and the regular icosahedron are shown in Figures 119 and 120, respectively. Notice, the outlines of both Petrie polygons form regular decagons.

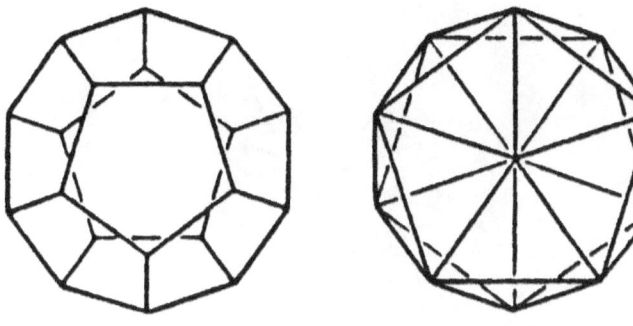

Figure 119   Figure 120

Two regular star polyhedra can be formed from the regular dodecahedron and the regular icosahedron. By extending the sides of a regular dodecahedron until the endpoints of these new extensions touch, a three-dimensional star can be formed. See Figure 121. A regular dodecahedron has twelve faces, so this new, regular star shape will have twelve vertices. Its vertices can form the twelve vertices of a regular icosahedron. This construction is called a star-dodecahedron.

Figure 121

By extending the sides of a regular icosahedron until the endpoints of these new extensions touch, a three-dimensional star can be formed. See Figure 122. A regular icosahedron has twenty faces, so this new, regular star shape will have twenty vertices. Its vertices can form the twenty vertices of a regular dodecahedron. This construction is called a star-icosahedron.

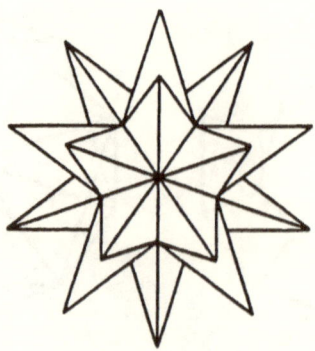

Figure 122

All regular solids can be inscribed, symmetrically, in a sphere. Tetra means four, octa means eight, dodeca means twelve, and icosa means twenty, terms all referring to the number of faces of the respective regular solid. All are terms, which come from the Greek language. The word cube comes from the Greek word kybos, which means "cube", or "vertebra" (*Webster's Seventh New Collegiate Dictionary*, 1963, s.v. "cube"). Hedra, also from Greek, means "seat" or "face", or "more at SIT" (<u>Ibid.</u>, s.v. "tetrahedron").

BIBLIOGRAPHY

Altieri, A. M. "Reflections on Fourth Dimension," *Mathematics Teacher*, XVIII (December, 1925), p. 494.

Asimov, Isaac. *Exploring the Earth & the Cosmos*. New York: Crown Publishers, Inc., 1982.

_____. *Asimov's Biographical Encyclopedia of Science and Technology, Second Revised Edition*. Garden City, N. Y.: Doubleday & Company, Inc., 1982.

Ball, W. W. Rouse. *Mathematical Recreations and Essays*. New York: The MacMillan Company, 1960.

Bankes, George, et al. *Eerdmans' Handbook to the World's Religions*. Grand Rapids, Michigan: Wm. B. Eerdmans Publishing Co., 1982.

Bell, E. T. *Men of Mathematics*. New York: Simon and Schuster, 1937.

Benson, Fred J. "Roads and Highways," *The New Encyclopaedia Britannica, Macropaedia* (15th ed.), 26, 1995.

Bergamini, David, and the Editors of Time-Life Books. *Mathematics*. New York: Time-Life Books, 1970.

Caminos, Ricardo A. "Pyramid," *Encyclopaedia Britannica*, 18, 1966.

Coxeter, H. S. M. *Introduction to Geometry*. New York: John Wiley and Sons, 1962.

_____. "The Golden Section, Phyllotaxis, and Wythoff's Game," *Scripta Mathematica*, 19 (June, 1953), p. 138.

Cundy, H. M., and Rollett, A. P. *Mathematical Models*. Oxford, England: Clarendon Press, 1957.

Drower, Margaret Stefana. "Egypt," *Encyclopaedia Britannica*, 8, 1966.

Euclid. *The Elements*, trans. by Sir Thomas L. Heath. New York: Dover Publications, Inc., 1956.

Eves, Howard. *An Introduction to the History of Mathematics*. New York: Holt, Rinehart, and Winston, 1976.

Frankel, Merrell. *Other Lands, Other Peoples*. Los Angeles: Los Angeles Unified School District, 1984.

Ghyka, Matila. *The Geometry of Art and Life*. New York: Dover Publications, Inc., 1977.

Gray, Eden. *The Tarot Revealed*. New York: The New American Library, Inc., 1960.

Hambidge, Jay. *The Diagonal*. New Haven, Conn.: Yale University Press, 1920.

_____. *The Parthenon and Other Greek Temples*. New Haven, Conn.: Yale University Press, 1924.

Heath, Sir Thomas Little, and Lloyd, Antony Charles. "Pythagoras and Pythagoreanism," *Encyclopaedia Britannica*, 18, 1966.

Herz, Annette. "The Fibonacci Numbers," *Journal of Undergraduate Mathematics*, 13 (March, 1982), p. 52.

Heyerdahl, Thor. *Early Man and the Ocean: A Search for the Beginnings of Navigation and Seaborne Civilizations*. Garden City, N. Y.: Doubleday & Company, Inc., 1978.

_____. *KON-TIKI: A Special Rand McNally Color Edition for Young People*. Chicago, New York: Rand McNally & Company, 1960.

_____. "The Voyage of Ra II," *National Geographic Magazine*, Vol. 139, No. 1 (January, 1971), pp. 44-71.

Hoffer, William. "Fibonacci Numbers," *Encyclopedia Science Supplement*, Grolier, Inc., 1977.

Holy Bible, King James Version.

Hudson, D. F. *New Testament Greek*. New York: David McKay Co., Inc., 1960.

Johnson, Chester. *Clocks and Watches*. New York: The Odyssey Press, 1964.

Jung, C. G. *Dreams*, trans. by R. F. C. Hull. Princeton, N. J.: Princeton University Press, 1974.

Kaplan, Stuart R. *Tarot Classic*. New York: U.S. Games Systems, Inc., 1972.

Kranzberg, Melvin. "Work and Employment," *The New Encyclopaedia Britannica, Macropaedia* (15th ed.), 29, 1995.

Kyhos, Gaither G. "A Traveler's Map of Spain and Portugal," *National Geographic Magazine*, Vol. 166, No. 4 (October, 1984), p. 460A.

Laurence, Theodor. *How the Tarot Speaks to Modern Man*. New York: Bell Publishing Company, 1972.

Macaulay, David. *Pyramid*. Boston: Houghton Mifflin Co., 1975.

MacQuitty, William. *Tutankhamun, The Last Journey*. New York: Crown Publishers, Inc., 1976.

Menninger, Karl. "Golden Section," *Encyclopaedia Britannica*, 10, 1972.

Moessel, Ernst. *Die Proportion In Antike und Mittelalter*. Munich: C. H. Beck, 1926.

Olds, C. D. *Continued Fractions*. New York: Random House, Inc., 1963.

Quilter, Jeffrey. "Prehistory in Peru," *Science*, 230 (November 22, 1985), pp. 931-932.

Redheffer, Ray. *Men of Modern Mathematics* (A history chart of mathematicians from 1000 to 1900). Somers, N. Y.: International Business Machines Corporation, 1966.

Stillwell, Richard. "Parthenon," *Encyclopaedia Britannica*, 17, 1966.

"The Alhambra; Spain," *Britannica Junior Encyclopaedia* (30th ed.), 2, 1966.

Thompson, John E. S. "Middle American Calendars," *Encyclopaedia Britannica*, 4, 1966.

Thompson, Sir D'Arcy Wentworth. *On Growth and Form*. Cambridge, England: University Press, 1942.

Thomsen, Dietrick E. "Calendric Reform in Yucatán," *Science News*, Vol. 126, No. 18 (November 3, 1984), pp. 282-3.

Tompkins, Peter. *Mysteries of the Mexican Pyramids*. New York: Harper & Row, 1976.

_____. *Secrets of the Great Pyramid*. New York: Harper & Row, 1971.

Tromp, Theresa. "The Fourth Dimension and Hyperspace," *Mathematics Teacher*, XIX (March, 1926), pp. 142-3.

Vance, Elbridge P. *An Introduction to Modern Mathematics*. Reading, Massachusetts: Addison-Wesley Publishing Company, 1968.

Voelkle, William M. "The Visconti-Sforza tarots," *Franco Maria Ricci*, No. 8 (January/February, 1985), p. 43.

Vorobyov, N. N. *The Fibonacci Numbers*. Boston: D. C. Heath and Co., 1963.

Wardlaw, C. W. *Essays on Form in Plants*. New York: Barnes and Noble, Inc., 1968.

Wenninger, Magnus. "Geodesic Domes by Euclidean Construction," *Mathematics Teacher*, 71 (October, 1978), pp. 582-587.

West, Beverly; Griesbach, Ellen; Taylor, Jerry; and Taylor, Louise. *The Prentice-Hall Encyclopedia of Mathematics*. Englewood Cliffs, New Jersey: Prentice-Hall, Inc., 1982.

Wilson, Carl Louis. "Flower," *Encyclopaedia Britannica*, 9, 1966.

Yates, Robert Carl. *Curves and Their Properties*. Washington, D.C.: The National Council of Teachers of Mathematics, 1974.

Recommended Reading:

Huntley, H. E. *The Divine Proportion: a Study in Mathematical Beauty*. New York: Dover Publications, Inc., 1970.

Lawlor, Robert. *Sacred Geometry, Philosophy and practice*. New York: The Crossroad Publishing Company, 1982.

www.ingramcontent.com/pod-product-compliance
Lightning Source LLC
Chambersburg PA
CBHW031959170526
45157CB00002B/466